# Acidophiles - Fundamentals and Applications

*Edited by Jianqiang Lin,*
*Linxu Chen and Jianqun Lin*

Published in London, United Kingdom

IntechOpen

*Supporting open minds since 2005*

Acidophiles - Fundamentals and Applications
http://dx.doi.org/10.5772/intechopen.87574
Edited by Jianqiang Lin, Linxu Chen and Jianqun Lin

Contributors
Rupesh K. Srivastava, Asha Bhardwaj, Leena Sapra, Bhupendra Verma, Lin-Xu Chen, Jianqun Lin, Jianqiang Lin, Lifeng Li, Zhaobao Wang, xianke chen, Xueyan Gao, Xin Pang, Juan Carlos Caicedo, Sonia Villamizar

Notice
Statements and opinions expressed in the chapters are these of the individual contributors and not necessarily those of the editors or publisher. No responsibility is accepted for the accuracy of information contained in the published chapters. The publisher assumes no responsibility for any damage or injury to persons or property arising out of the use of any materials, instructions, methods or ideas contained in the book.

First published in London, United Kingdom, 2021 by IntechOpen
IntechOpen is the global imprint of INTECHOPEN LIMITED, registered in England and Wales, registration number: 11086078, 5 Princes Gate Court, London, SW7 2QJ, United Kingdom
Printed in Croatia

British Library Cataloguing-in-Publication Data
A catalogue record for this book is available from the British Library

Additional hard and PDF copies can be obtained from orders@intechopen.com

Acidophiles - Fundamentals and Applications
Edited by Jianqiang Lin, Linxu Chen and Jianqun Lin
p. cm.
Print ISBN 978-1-83969-279-6
Online ISBN 978-1-83969-280-2
eBook (PDF) ISBN 978-1-83969-281-9

We are IntechOpen,
the world's leading publisher of
Open Access books
Built by scientists, for scientists

## 5,500+
Open access books available

## 136,000+
International authors and editors

## 170M+
Downloads

Our authors are among the

## 156
Countries delivered to

## Top 1%
most cited scientists

## 12.2%
Contributors from top 500 universities

CLARIVATE ANALYTICS
BOOK
CITATION
INDEX
INDEXED

WEB OF SCIENCE™

Selection of our books indexed in the Book Citation Index (BKCI)
in Web of Science Core Collection™

Interested in publishing with us?
Contact book.department@intechopen.com

# Meet the editors

Dr. Jianqiang Lin has been a professor at the State Key Laboratory of Microbial Technology (SKLMT), Shandong University, China, since 2003. He obtained a Ph.D. in 1998 under a China-Japan government cooperation program on Ph.D. education and completed his thesis research at Osaka University, Japan. In 2000, he was a visiting researcher at Inha University, Korea. His research interests include molecular biology of acidophiles, microbial technology, and bioprocess bioengineering.

Dr. Linxu Chen has a Ph.D. in Microbiology from the State Key Laboratory of Microbial Technology (SKLMT), Shandong University, China. He completed his postdoctoral research at Louisiana State University, USA. His research interests include gene transfer systems and gene-editing technology of chemoautotrophic acidophilic bacteria, metabolic and adaptive mechanisms of bioleaching bacteria, and genetic engineering and synthetic biology of bioleaching bacteria.

Dr. Jianqun Lin is Director of the Shandong Society for Microbiology, China. He obtained a Ph.D. from the Chinese Academy of Sciences in 1997. He was a visiting researcher at Osaka University, Japan, in 1997–1998, and completed his postdoctoral research at the Uniformed Services University of Health Science, USA. His research interests include biological resources development of acidophiles in acid mine environments, bacteria–mineral interaction and the sulfide ore bioleaching process, and the development of high-efficiency bioleaching techniques.

# Contents

# Preface

Acidophiles, as an important class of extremophiles, have attracted attention for their scientific significance and application value. Acidophiles are microorganisms that thrive in acidic conditions and can be found in Bacteria, Archaea, and Eukarya. Acidophiles can be identified in both terrestrial and marine environments, including acid mine drainage (AMD) from metal and coal mines, iron-sulfur mineral mines, hot springs, and sediments. In these acidic ecosystems, acidophiles could participate in the element cycles (sulfur, iron, etc.) and promote the generation of the acid environment and formation of the acid micro-ecosystem. Acidophiles can also be found in the human body. Research on acidophiles can help us understand the diversity, adaptation, and functions of these microorganisms and contribute to the development and application of new biotechnologies for resolving problems of resource exploitation, pollution, and human disease. This book provides breakthroughs and insights into the research on acidophiles. Chapter 1 introduces acidophiles and their important physiological characteristics and industrial applications. Chapter 2 discusses the two-component signal systems in the regulation of sulfur and ferrous iron oxidation in acidophilic bacteria. Chapter 3 focuses on the acid-resistance adaptation mechanisms in acidophiles. Chapter 4 introduces the quorum sensing system (QS)-mediated communication mode in chemoautotrophic acidophilic bacteria and the function of QS in the regulation of bacterial sulfur oxidation, ferrous iron oxidation, and mineral oxidation. Chapter 5 discusses the application of QS and QS signals on bioleaching industries. The final chapter deals with acidophiles in human health, including the immunomodulatory potential of *Lactobacillus acidophilus* and its potential role as a therapeutic for the management of inflammation-induced bone disorders.

**Jianqiang Lin, Linxu Chen and Jianqun Lin**
State Key Laboratory of Microbial Technology,
Shandong University,
Qingdao, China

# Introductory Chapter: The Important Physiological Characteristics and Industrial Applications of Acidophiles

*Linxu Chen, Jianqun Lin and Jianqiang Lin*

## 1. The definition of acidophiles

Acidophiles are an important category of extremophiles that are defined by the environmental conditions in which they grow optimally. Acidophile is a broad definition that organisms can grow preferentially in environments with a pH at below 6. In 2007, Johnson proposed a generally accepted classification standard according to the optimal pH. The organisms with optimal pH at 3 or below are classified as extreme acidophiles, and those with an optimal pH of 3–5 are moderate acidophiles [1]. Although some organisms can grow at a pH lower than 5, they are recognized as acid-tolerant species because of their pH optima above 5. The research history of acidophiles started in the discovery of a sulfur-oxidizing bacteria isolated from a compost sample mixed with sulfur, rock phosphate, and soil by Waksman and Joffe in 1922 [2]. This bacterium has an optimal growth pH at 2.0–2.8 and is a strict auto-troph that obtains energy by oxidizing inorganic sulfur substances (elemental sulfur, thiosulfate, and hydrogen sulfide). This bacterium was named as *Thiobacillus thiooxidans* by Waksman and Joffe, and later was reclassified as *Acidithiobacillus thiooxidans* by Kelly and Wood in 2000 [3]. With the development of microbiology and gene sequencing technology, more and more acidophiles have been discovered, identified, and sequenced. Until now, the most acidophilic organisms are from an archaeal genus of *Picrophilus*, firstly isolated from acidic hot springs and dry hot soil in Hokkaido in Japan [4]. Members in *Picrophilus* have optimal pH at 0.7 and the ability to grow at a pH of 0. Moreover, acidophiles are involved not in the domains of *Bacteria* and *Archaea*, but also in the *Eukarya* domains, such as some acidophilic fungi, algae, and yeast distributed in the acid mine environments.

## 2. The typical acidophilic bacteria and the applications of acidophiles

*Acidithiobacillus* is a kind of extensive research and wide application of gram-negative acidophiles. Members in this genus are broadly existed in the sulfur-containing acidic environments on land or in the sea, such as acid mine drainage (AMD), iron–sulfur mineral mines, hot springs, and sediments [5–10]. *Acidithiobacillus* spp., as the important sulfur and iron-oxidizers, participate in the element cycles of sulfur and iron, and promote the acid environment genera-tion and acid microecosystem formation. All *Acidithiobacillus* strains have the capability of oxidizing various reduced inorganic sulfur compounds (RISCs)

| Trait | A. ferrooxidans | Acidithiobacillus ferrivorans | A. ferriphilus | A. ferridurans | Acidithiobacillus thiooxidans | Acidithiobacillus caldus | A. albertensis |
|---|---|---|---|---|---|---|---|
| Gram stain | – | – | – | – | – | – | – |
| Cell size (μm) | 1.0 × 0.5 | 2.4 × 0.5 | 1–2 | 1–2 | 1.0–2.0 × 0.5 | 1.2–1.9 × 0.7 | 1–2 × 0.4–0.6 |
| Motility | +/– | + | + | + | + | + | + |
| Growth pH (optimum) | 1.3–4.5 (2.0–2.5) | 1.9–3.4 (2.5) | 1.5– (2.0) | 1.4–3.0 (2.1) | 0.5–5.5 (2.0–3.0) | 1.0–3.5 (2.0–2.5) | 0.5–6.0 (3.5–4.0) |
| Growth T/°C (optimum) | 10–37 (30–35) | 4–37 (28–33) | 5–33 (30) | 10–37 (29) | 10–37 (28–30) | 32–52 (40–45) | 10–40 (25–30) |
| Oxidation of $S^0$, $S_4O_6^{2-}$, $S_2O_3^{2-}$ | + | + | + | + | + | + | + |
| Oxidation of $Fe^{2+}$ | + | + | + | + | – | – | – |
| Growth on sulfide minerals | + | + | + | + | – | – | – |
| Growth on hydrogen | + | (+) | – | + | – | + | NR |
| Anaerobic growth with $Fe^{3+}$ | + | + | + | + | – | – | – |
| $N_2$ fixation | + | + | NR | NR | – | – | – |
| Mol% G + C | 58–59 | 55–56 | 57.4 | 58.4 | 52 | 63–64 | 61.5 |
| Thiosulfate-metabolic pathways | TSD enzyme; S4I pathway. | Sox system; TSD enzyme; S4I pathway. | NR | TSD enzyme; S4I pathway. | Sox system; S4I pathway. | Sox system; S4I pathway. | Sox system; S4I pathway. |

**Table 1.**
*Taxonomic traits of species in the genus of Acidithiobacillus.*

+, positive; –, negative; +/–, the positive or negative result from different reports; (+), some strains have the ability to oxidize hydrogen; NR, not reported; Tm, temperature.

and elemental sulfur, and some of them also have ferrous iron oxidation ability [11]. By the oxidation of sulfur and ferrous, *Acidithiobacillus* spp. obtains electrons to generate the bioenergy (ATP) and reducing power (NADH/NADPH) to fix carbon dioxide for autotrophic growth. More and more species have been identified based on their physiological characters and 16S rRNA gene sequences (**Table 1**) [2, 12–16]. Species in *Acidithiobacillus* can be divided into two groups according to their energy-substrates: the sulfur-oxidizing only species, including *A. thiooxidans*, *Acidithiobacillus caldus* and *A. albertensis*, and the sulfur and ferrous-oxidizing species, including *A. ferrooxidans*, *Acidithiobacillus ferrivorans*, *A. ferriphilus*, and *A. ferridurans* (**Table 1**).

*Acidithiobacillus* spp. and other chemoautotrophic acidophilic bacteria have an important application in bioleaching. The bioleaching technology is originated from the biohydrometallurgy industry, and has become a great potential and broad-prospects in non-ferrous metal extraction (golden, silver, copper et al.) from various sulfide ores. *Acidithiobacillus* spp. have the remarkable capabilities of metabolizing the sulfur and iron in ores and adapting to extremely acidic environments, thus they have become the most active and preponderant bacteria used in biomining [17, 18]. *A. ferrooxidans*, *A. thiooxidans*, and *A. caldus* are the wide used ore leaching species in biomining for mineral extraction from ores [19, 20]. In recent years, based on their abilities to produce acid and heavy leaching metals, *Acidithiobacillus* spp. have been used from biohydrometallurgy to the treatment of wastes containing heavy metals, such as sewage sludge, spent household batteries, mine tailings, and printed circuit boards [21–25]. Moreover, these bacteria have been widely studied in microbial desulfurization of coal and gas [26–28]. In a word, the great application values of *Acidithiobacillus* spp. have been exploited from the biohydrometallurgy industry to the environmental pollution treatments.

## 3. The physiological feature of chemoautotrophic acidophiles

Sulfur oxidation is a characteristic physiological feature for many acidophilic microorganisms and is an important biochemical process that promotes the generation of the acid environment and the formation of acidophilic microbial communities. *Acidithiobacillus* spp., as the first-discovered and the most widespread used acidophile, has been attracted extensive attention and has been used as model sulfur-oxidizing bacteria to research microbial sulfur metabolism [11, 29–40]. The oxidation states of element sulfur are range from −2 to +6, resulting in different kinds of RISCs (tetrathionate ($S_4O_6^{2-}$), thiosulfate ($S_2O_3^{2-}$), sulfite ($SO_3^{2-}$), sulfide ($S^{2-}$) et al.), and elemental sulfur ($S^0$). Thus, many microbes, particularly autotrophic sulfur-oxidizing microbes, have evolved a variety of enzymes and proteins participating in the oxidation of RISCs and $S^0$. Research shows *Acidithiobacillus* spp. have a high-efficient and sophisticated sulfur-metabolizing network that could oxidize RISCs and $S^0$ to sulfate. Based on metabolic substrates, the sulfur-metabolic enzymes in *Acidithiobacillus* spp. could be categorized as elemental sulfur oxidation enzymes, enzymes in thiosulfate oxidation pathways, sulfide oxidation enzymes, and sulfite oxidation enzymes. These enzymes work cooperatively in different cellular compartments to oxidize the RISCs and $S^0$ to the final product sulfate (**Figures 1** and **2**) [11]. As shown in **Figure 1**, the extracellular elemental sulfur ($S_8$) oxidation in *A. caldus* starts from the activation and transportation of $S_8$ by special outer-membrane proteins (OMP), generating the persulfide sulfane sulfur in the periplasm; then the persulfide sulfane sulfur is oxidized to sulfite that can directly enter the sulfur oxidizing enzyme (Sox) system or form $S_2O_3^{2-}$ via a nonenzymatic reaction; the generated thiosulfate is then metabolized by the truncated Sox pathway or catalyzed by thiosulfate:quinol oxidoreductase (TQO or DoxDA) to

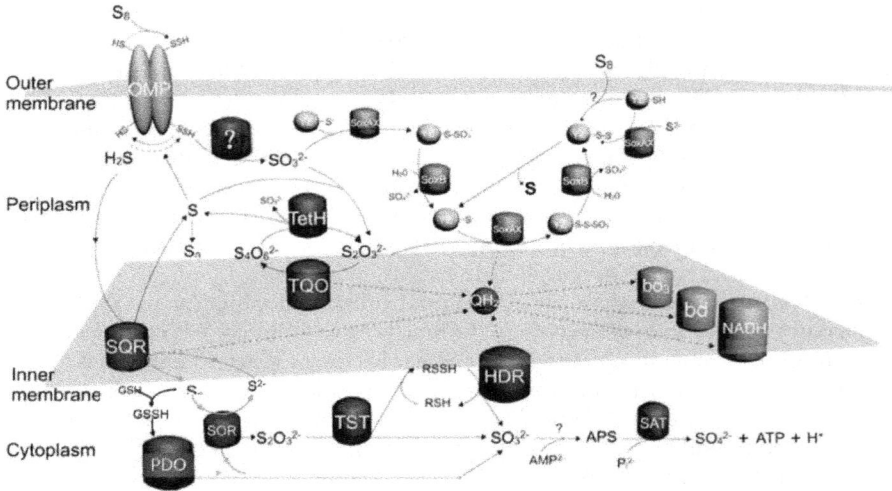

**Figure 1.**
*The model of sulfur oxidation in Acidithiobacillus caldus. OMP, outer-membrane proteins; TQO, thiosulfate quinone oxidoreductase; TetH, tetrathionate hydrolase; SQR, sulfide:Quinone oxidoreductase; PDO, persulfide dioxygenase; SOR, sulfur oxygenase reductase; TST, rhodanese; HDR, Hdr-like complex; SAT, ATP sulfurylase; bd, $bo_3$, terminal oxidases; $QH2$, quinol pool; NADH, NADH dehydrogenase complex I.*

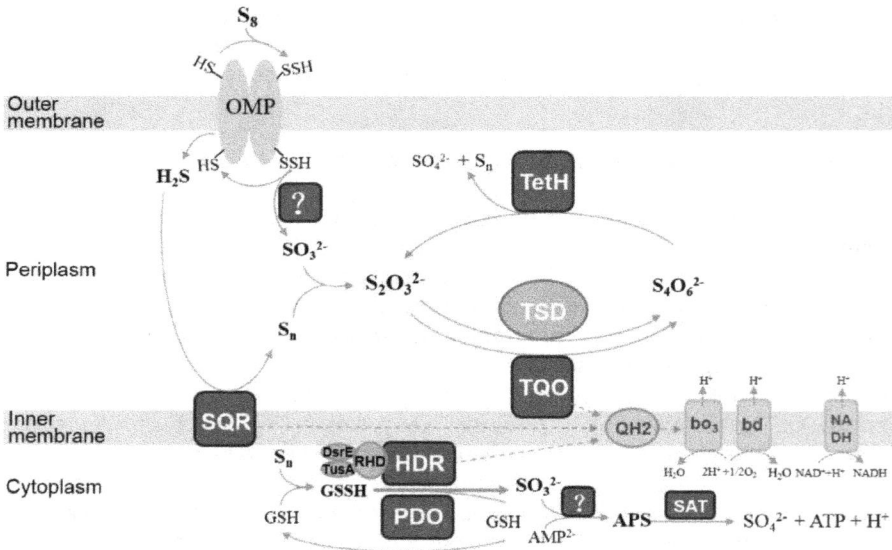

**Figure 2.**
*The model of sulfur oxidation in A. ferrooxidans. OMP, outer-membrane proteins; TQO, thiosulfate quinone oxidoreductase; TSD, thiosulfate dehydrogenase; TetH, tetrathionate hydrolase; SQR, sulfide:Quinone oxidoreductase; PDO, persulfide dioxygenase; HDR, Hdr-like complex; SAT, ATP sulfurylase; bd, $bo_3$, terminal oxidases; $QH2$, quinol pool; NADH, NADH dehydrogenase complex I.*

generate $S_4O_6^{2-}$; $S_4O_6^{2-}$ is further hydrolyzed by tetrathionate hydrolase (TetH); the $H_2S$ produced during the activation of $S_8$ can be oxidized by sulfide:quinone oxidore-ductase (SQR) located in the inner membrane; the periplasmic elemental sulfur (Sn) produced from Sox pathway, tetrathionate hydrolysis and sulfide oxidation, could be re-activated at the outer membrane region, or be mobilized into the cytoplasm where Sn could be used by cytoplasmic elemental sulfur oxidation enzyme persulfide dioxygenase (PDO) and Sulfur oxygenase reductase (SOR); the metabolites from the reaction of PDO and SOR could be utilized by cytoplasmic sulfur-metabolic enzymes, including the $S_2O_3^{2-}$ metabolism via by rhodanese (TST) and the Hdr-like complex (HDR), the degradation of $SO_3^{2-}$ via the APS pathway and the oxidation of $S^{2-}$ by SQR. During the sulfur metabolic process, the periplasmic sulfur-oxidizing pathways (Sox and TetH) are responsible for electron acquisition, thus they are important for the sulfur metabolism in *A. caldus*. Different from '*A. caldus*' like sulfur metabolism network, some sulfur-oxidizers, such as *A. ferrooxidans*, did not have the Sox pathway, but rather a thiosulfate dehydrogenase (TSD) (**Figure 2**). Interestingly, *A. ferrivorans* possesses both Sox system and TSD enzyme (**Table 1**). The proposal of sulfur metabolism models provides new knowledge and insights in understanding the metabolism and adaptation mechanisms of acidophilic sulfur-oxidizing microorganisms in extreme environments.

## 4. The significance of studying and understanding acidophiles

Acidophiles, as important extremophiles, have presented important scientific significance and industrial application values. Researches on acidophiles do not only help us understand the diversity and adaptation of life on earth, but also be conducive in developing various new biotechnologies to resolve the problems of resource exploitation, pollution treatment, and human health. This book provides some new breakthroughs and insights on the researches of acidophiles: the two-component system (TCS) in the regulation of sulfur metabolic process; the adaptation mechanisms of acidophiles to low pH; the regulation mechanism and the application strategy of quorum sensing in bioleaching bacteria; *Lactobacillus acidophilus* and its application in the human health.

Author details

Linxu Chen*, Jianqun Lin and Jianqiang Lin*
State Key Laboratory of Microbial Technology, Shandong University,
Qingdao, China

*Address all correspondence to: linxuchen@sdu.edu.cn
and jianqianglin@sdu.edu.cn

IntechOpen

# References

[1] Johnson JD. Physiology and ecology of acidophilic microorganisms. In: Gerday C, Glansdorff N, editors. Physiology and Biochemistry of Extremophiles. Washington, DC, USA: American Society of Microbiology Press; 2007. pp. 255-270

[2] Waksman SA, Joffe J. Microorganisms concerned in the oxidation of sulfur in the soil: II. *Thiooxidans*, a new sulfur-oxidizing organism isolated from the soil. Journal of Bacteriology. 1922;**7**(2):239

[3] Kelly DP, Wood AP. Reclassification of some species of *Thiobacillus* to the newly designated genera *Acidithiobacillus* gen. nov., *Halothiobacillus* gen. nov. and *Thermithiobacillus* gen. nov. International Journal of Systematic and Evolutionary Microbiology. 2000;**50**(2):511-516

[4] Schleper C et al. *Picrophilus* gen. nov., fam. nov.: A novel aerobic, heterotrophic, thermoacidophilic genus and family comprising archaea capable of growth around pH 0. Journal of Bacteriology. 1995;**177**(24):7050-7059

[5] Harrison AP Jr. The acidophilic thiobacilli and other acidophilic bacteria that share their habitat. Annual Review of Microbiology. 1984;**38**(1):265-292

[6] Schrenk MO et al. Distribution of *Thiobacillus ferrooxidans* and *Leptospirillum ferrooxidans*: Implications for generation of acid mine drainage. Science. 1998;**279**(5356):1519-1522

[7] Jones DS et al. Community genomic analysis of an extremely acidophilic sulfur-oxidizing biofilm. The ISME Journal. 2012;**6**(1):158

[8] Hua Z-S et al. Ecological roles of dominant and rare prokaryotes in acid mine drainage revealed by metagenomics and meta transcriptomics. The ISME Journal. 2015;**9**(6):1280

[9] Sharmin S et al. Characterization of a novel thiosulfate dehydrogenase from a marine acidophilic sulfur-oxidizing bacterium, *Acidithiobacillus thiooxidans* strain SH. Bioscience, Biotechnology, and Biochemistry. 2016;**80**(2):273-278

[10] Nuñez H et al. Molecular systematics of the genus *Acidithiobacillus*: Insights into the phylogenetic structure and diversification of the taxon. Frontiers in Microbiology. 2017;**8**:30

[11] Wang R et al. Sulfur oxidation in the acidophilic autotrophic *Acidithiobacillus* spp. Frontiers in Microbiology. 2019;**9**:3290

[12] Temple KL, Colmer AR. The autotrophic oxidation of iron by a new bacterium: *Thiobacillus ferrooxidans*. Journal of Bacteriology. 1951;**62**(5):605

[13] Hallberg KB, Lindström EB. Characterization of *Thiobacillus caldus* sp. nov., a moderately thermophilic acidophile. Microbiology. 1994;**140**(12): 3451-3456

[14] Xia J-L et al. A new strain *Acidithiobacillus* albertensis BY-05 for bioleaching of metal sulfides ores. Transactions of Nonferrous Metals Society of China. 2007;**17**(1):168-175

[15] Liljeqvist M, Rzhepishevska OI, Dopson M. Gene identification and substrate regulation provides insights into sulfur accumulation during bioleaching with the psychrotolerant *Acidithiobacillus ferrivorans*. Applied and Environmental Microbiology. 2013;**79**(3):951-957

[16] Falagán C, Johnson DB. *Acidithiobacillus ferriphilus* sp. nov., a facultatively anaerobic iron-and

sulfur-metabolizing extreme acidophile. International Journal of Systematic and Evolutionary Microbiology. 2016;**66**(1): 206-211

[17] Rawlings D, Tributsch H, Hansford G. Reasons why 'Leptospirillum'-like species rather than *Thiobacillus ferrooxidans* are the dominant iron-oxidizing bacteria in many commercial processes for the biooxidation of pyrite and related ores. Microbiology. 1999;**145**(1):5-13

[18] Rawlings DE. Characteristics and adaptability of iron-and sulfur-oxidizing microorganisms used for the recovery of metals from minerals and their concentrates. Microbial Cell Factories. 2005;**4**(1):13

[19] Valdés J et al. *Acidithiobacillus ferrooxidans* metabolism: From genome sequence to industrial applications. BMC Genomics. 2008;**9**(1):597

[20] Valdés J et al. Comparative genome analysis of *Acidithiobacillus ferrooxidans*, *A. thiooxidans* and *A. caldus*: Insights into their metabolism and ecophysiology. Hydrometallurgy. 2008;**94**(1-4):180-184

[21] Pathak A, Dastidar M, Sreekrishnan T. Bioleaching of heavy metals from sewage sludge. Journal of Environmental Management. 2009;**90**(8):2343-2353

[22] Bayat B, Sari B. Comparative evaluation of microbial and chemical leaching processes for heavy metal removal from dewatered metal plating sludge. Journal of Hazardous Materials. 2010;**174**(1-3):763-769

[23] Arshadi M, Mousavi S. Simultaneous recovery of Ni and Cu from computer-printed circuit boards using bioleaching: Statistical evaluation and optimization. Bioresource Technology. 2014;**174**:233-242

[24] Nguyen VK et al. Bioleaching of arsenic and heavy metals from mine tailings by pure and mixed cultures of *Acidithiobacillus* spp. Journal of Industrial and Engineering Chemistry. 2015;**21**:451-458

[25] Rastegar S et al. Bioleaching of V, Ni, and Cu from residual produced in oil fired furnaces using *Acidithiobacillus ferrooxidans*. Hydrometallurgy. 2015;**157**:50-59

[26] Azizan A, Najafpour G, Harun A. Microbial desulfurization of coal with mixed cultures of *Thiobacillus ferrooxidans* and *Thiobacillus thiooxidans*. In: Proceedings of the 2nd International Conference on Advances in Strategic Technologies-ICAST; 2000

[27] Charnnok B et al. Oxidation of hydrogen sulfide in biogas using dissolved oxygen in the extreme acidic biofiltration operation. Bioresource Technology. 2013;**131**:492-499

[28] He H et al. Biodesulfurization of coal with *Acidithiobacillus caldus* and analysis of the interfacial interaction between cells and pyrite. Fuel Processing Technology. 2012;**101**:73-77

[29] Suzuki I, Werkman C. Glutathione and sulfur oxidation by *Thiobacillus thiooxidans*. Proceedings of the National Academy of Sciences. 1959;**45**(2): 239-244

[30] London J. Cytochrome in *Thiobacillus thiooxidans*. Science. 1963;**140**(3565):409-410

[31] London J, Rittenberg S. Path of sulfur in sulfide and thiosulfate oxidation by thiobacilli. Proceedings of the National Academy of Sciences. 1964;**52**(5):1183-1190

[32] Suzuki I, Chan C, Takeuchi T. Oxidation of elemental sulfur to sulfite by *Thiobacillus thiooxidans* cells. Applied and Environmental Microbiology. 1992;**58**(11):3767-3769

[33] Hallberg KB, Dopson M, Lindström EB. Reduced sulfur compound oxidation by *Thiobacillus caldus*. Journal of Bacteriology. 1996;**178**(1):6-11

[34] Quatrini R et al. Extending the models for iron and sulfur oxidation in the extreme acidophile *Acidithiobacillus ferrooxidans*. BMC Genomics. 2009; **10**(1):394

[35] Chen L et al. *Acidithiobacillus caldus* sulfur oxidation model based on transcriptome analysis between the wild type and sulfur oxygenase reductase defective mutant. PLoS One. 2012;7(9): e39470

[36] Yin H et al. Whole-genome sequencing reveals novel insights into sulfur oxidation in the extremophile *Acidithiobacillus thiooxidans*. BMC Microbiology. 2014;**14**(1):179

[37] Li L-F et al. The σ 54-dependent two-component system regulating sulfur oxidization (Sox) system in *Acidithiobacillus caldus* and some chemolithotrophic bacteria. Applied Microbiology and Biotechnology. 2017;**101**(5):2079-2092

[38] Wang Z-B et al. The two-component system RsrS-RsrR regulates the tetrathionate intermediate pathway for thiosulfate oxidation in *Acidithiobacillus caldus*. Frontiers in Microbiology. 2016;7:1755

[39] Wu W et al. Discovery of a new subgroup of sulfur dioxygenases and characterization of sulfur dioxygenases in the sulfur metabolic network of *Acidithiobacillus caldus*. PLoS One. 2017;**12**(9):e0183668

[40] Yu Y et al. Construction and characterization of tetH overexpression and knockout strains of *Acidithiobacillus ferrooxidans*. Journal of Bacteriology. 2014;**196**(12):2255-2264

Chapter 2

# Two-Component Systems in the Regulation of Sulfur and Ferrous Iron Oxidation in Acidophilic Bacteria

*Lifeng Li and Zhaobao Wang*

## Abstract

The two-component system (TCS) is a regulatory system composed of a sensor histidine kinase (HK) and a cytoplasmic response regulator (RR), which participates in the bacterial adaptation to external stimuli. Sulfur oxidation and ferrous iron oxidation are basic energy metabolism systems for chemoautotrophic acidophilic bacteria in acid mine environments. Understanding how these bacteria perceive and respond to complex environmental stimuli offers insights into oxidization mechanisms and the potential for improved applications. In this chapter, we summarized the TCSs involved in the regulation of sulfur and ferrous iron metabolic pathways in these acidophilic bacteria. In particular, we examined the role and molecular mechanism of these TCSs in the regulation of iron and sulfur oxidation in *Acidithiobacillus* spp.. Moreover, research perspectives on TCSs in acidophilic bacteria are discussed in this section.

**Keywords:** *Acidithiobacillus*, two-component system, ferrous iron oxidation, sulfur oxidation, transcriptional regulation

## 1. Introduction

*Acidithiobacillus* genus is composed of high acid-tolerance chemolithotrophic bacteria that can oxidize various reduced inorganic sulfur compounds (RISCs) and ferrous iron to obtain electrons for carbon dioxide fixation and energy production [1]. The composition and comparison of the members in this genus has been reviewed [2]. As reported, the bacteria can be classified into two groups based on their energy resources: the sulfur-oxidizing-only species and the sulfur- and ferrous- oxidizing species [3]. Sulfur-oxidizing-only bacteria include *Acidithiobacillus caldus*, *Acidithiobacillus thiooxidans* and *Acidithiobacillus albertensis*, whereas sulfur- and ferrous oxidizing bacteria include *Acidithiobacillus ferrooxidans*, *Acidithiobacillus ferrivorans*, *Acidithiobacillus ferriphilus*, and *Acidithiobacillus ferridurans*. These bacteria are widespread in the bioleaching heap and acid mine drainage water environments and play critical roles in bioleaching and wastewater treatment [3–6]. Sulfur and iron oxidation capacities are critical physiological features of these bacteria, which are also the basis for their applications. The oxidation of reduced inorganic sulfur compounds can dissolve ore and produce sulfuric acid,

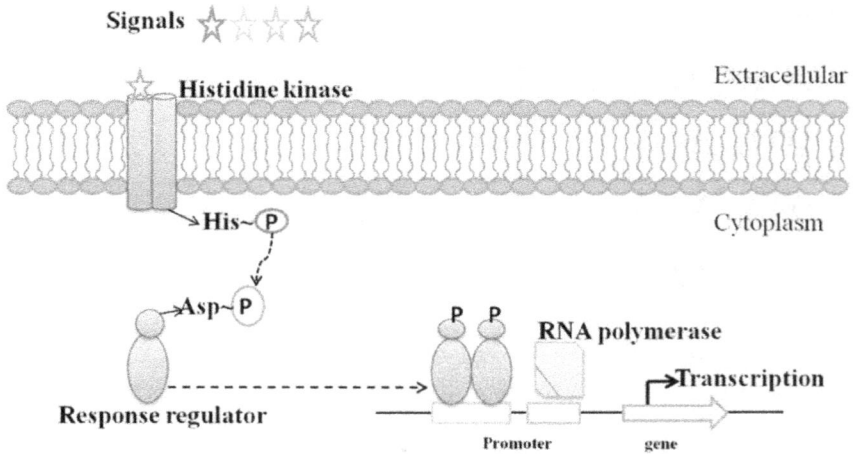

**Figure 1.**
*Two-component system regulation mechanism.*

whereas the oxidation of ferrous iron (Fe II) produces ferric iron (Fe III), in which sulfuric acid and ferric iron products can attack minerals, releasing metal ions [7]. Sulfur metabolism and iron oxidation are complicated and various metabolic genes are involved. Thus, the regulation and mechanism of the sulfur and iron oxidation in *Acidithiobacillus* spp. have drawn increasing attention.

Sensing and responding to environmental stimuli is necessary for bacteria to adjust the expression of related genes and adapt to changing habitats. The two-component systems (TCSs) are the most widespread regulation system in bacteria [8]. The TCS is mainly composed of two proteins, histidine kinase (HK) and their cognate response regulator (RR) (**Figure 1**). Histidine kinase is a membrane protein that can sense extracellular signals and autophosphorylate its histidine. The phosphorylated HK can transfer the phosphoryl group to its cognate RR protein leading to the phosphorylation of the RR protein at the aspartate residue (Asp) and the activation of RR protein. The activated RR protein is able to change its conformation by dimerization or multimerization and regulates the expression of its target genes. In general, the RR protein can regulate gene transcription by binding to specific sequences in the promoter region of related genes located upstream of the RNA polymerase binding region.

Although not completely understood, the study of molecular regulation mechanisms in acidophilic bacteria has recently been progressing. In this chapter, we discuss the occurrence of the TCS in these bacteria, the regulation mechanism of sulfur and iron oxidation, and the future prospects in the TCS regulation research.

## 2. Discovery of two-component system in acidophilic bacteria

The occurrence of the TCSs in the acidophilic bacteria was compared among different species on basis of the reported TspS-TspR, RsrS-RsrR, and RegB-RegA two-component systems [7, 9, 10] (**Figure 2**). The sulfur oxidization (Sox) system is a critical sulfur oxidization pathway of chemotrophic sulfur-oxidizing bacteria, and the regulation of the Sox system in *A. caldus* by the TspS-TspR two-component system has been reported [10]. Meanwhile, genome sequences were used to

**A. TCS in Sox pathway**

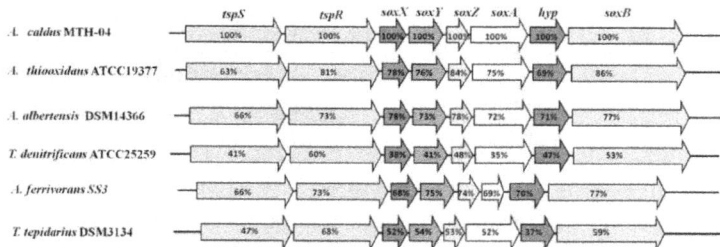

**B. TCS in S₄I pathway**

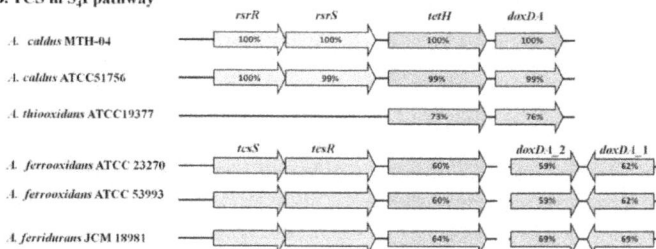

**C. TCS in iron oxdation pathway**

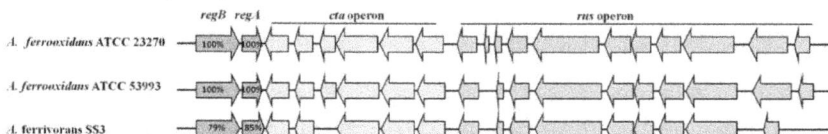

**Figure 2.**
*Distribution of two-component system in acidophilic bacteria. The identities of corresponding protein were indicated by the percentage values with the first line of each part set as 100%. Accession numbers (GenBank) for proteins in Sox pathway are as follows, A. caldus MTH-04, sox (A5904_11270–11305); A. thiooxidans ATCC19377, sox (ATHIO_RS0101665-RS0101630); A. albertensis DSM 14366, sox (BLW97_RS11430-RS11465); A. ferrivorans SS3, sox (Acife_2487–2494). Accession numbers (GenBank) for proteins in S₄I pathway are as follows, A. caldus MTH-04, RsrR (ANJ65973.1), RsrS (ANJ65974.1), TetH (OAN03451.1), DoxDA (OAN03452.1) (GenBank: MK165448); A. caldus ATCC 51756, RsrR (ABP38227.1), RsrS (ABP38226.1), TetH (ABP38225.1), DoxDA (ABP38224.1); A. thiooxidans ATCC 19377, TetH (WP_029316048.1), DoxDA (WP_010638552.1); Acidithiobacillus ferrooxidans ATCC 23270, TcsS (ACK79489.1), TcsR (ACK79259.1), TetH (ACK80599.1), DoxDA_2 (ACK79881.1), DoxDA_1 (ACK78481.1); A. ferrooxidans ATCC 53993, TcsS (ACH82290.1), TscR (ACH82291.1), TetH (ACH82292.1), DoxDA_2(ACH82311.1), DoxDA_1(ACH82307.1); Acidithiobacillus ferridurans JCM 18981, TcsS (BBF65177.1), TcsR (BBF65176.1), TetH (BBF65175.1), DoxDA_2 (BBF65156.1), DoxDA_1 (BBF65160.1). Accession numbers (GenBank) for proteins in iron oxidation pathway are as follows, Acidithiobacillus ferrooxidans ATCC 23270, AFE_RS14375-AFE_RS16285; A. ferrooxidans ATCC 53993, LFERR_RS13505-LFERR_RS15550; A. ferrivorans SS3, ACIFE_RS08920-ACIFE_RS09010.*

compare the occurrence of TCS similar to that of TspS-TspR in the *Acidithiobacillus* spp.. *tspS-tspR-sox* gene clusters were found in all the sulfur-oxidizing bacteria (*Acidithiobacillus caldus*, *Acidithiobacillus thiooxidans* and *Acidithiobacillus albertensis*), and in one type of the sulfur- and ferrous- oxidizing bacteria, *Acidithiobacillus ferrivorans*. The amino acid similarity of the TspS/TspR ranged from 63% to 81%.

The S₄I pathway is also an important thiosulfate oxidization pathway composed of tetrathionate hydrolase (TetH) and thiosulfate: quinone oxidoreductase (DoxDA), and its regulation by the RsrS-RsrR system was reported [9, 11]. However, a similar distribution of this gene cluster was only found in *A. caldus*. No regulatory system was found in *A. thiooxidans*. In contrast, a different kind of TCS was found before the *tetH* gene in *Acidithiobacillus ferrooxidans* and *Acidithiobacillus ferridurans* with two *doxDA* genes separated in a different gene cluster.

The RegB-RegA is a well-studied global redox responding regulatory system in *A. ferrooxidans*, and plays roles in the iron and sulfur oxidization regulation [12, 13]. RegB-RegA located in the *cta* and *rus* operon, which was composed of genes in the biogenesis of aa3 type oxidase and iron oxidation pathway. Similarly, the *regB-regA-cta-rus* cluster was only found in the sulfur- and ferrous- oxidizing bacteria *A. ferrooxidans* and *A. ferrivorans* with high identity.

Hence, the TCSs are widespread in the sulfur and iron oxidization bacteria, while different distributions are revealed by bioinformatics analysis and different regulation mechanism maybe adapted, which deserves further studies.

## 3. Roles of two-component system in sulfur oxidation

Gene transcription is a fundamental process in bacteria, which is carried out by multi-subunit RNA polymerase (RNAP). σ factors determine transcription specificity by recognizing specific promoter sequences. Bacterial σ factors can be divided into two distinct classes: $\sigma^{70}$ and $\sigma^{54}$ [14]. $\sigma^{70}$ recognizes the consensus −10 and − 35 regions and recruits RNAP to a specific promoter region to initiate gene transcription [15]. $\sigma^{70}$ controls transcription of most housekeeping genes, whereas $\sigma^{54}$ regulates the genes involved in nitrogen assimilation [16], phage shock response [17], infection [18], and other cellular stresses [19, 20]. $\sigma^{54}$ recognizes distinct sequences in the −12 (GC) and − 24 (GG) regions of the promoter. The requirement of the bacterial enhancer binding proteins (bEBPs) is a remarkable feature of $\sigma^{54}$-dependent transcription initiation [20]. Accordingly, two kinds of transcription regulation were reported in acidophilic bacteria (**Figure 3**).

The *rsrS-rsrR-tetH-doxDA* gene cluster in *A. caldus* was reported in 2007 [11]. The genes in this cluster were proven to be cotranscribed using RT-PCR (Reverse transcription PCR), and the results of quantitative PCR and Western blot indicated that the gene cluster was tetrathionate induced. The promoter before *tetH* gene was mapped by primer extension. The verification of the regulation role and mechanism of the S$_4$I pathway by the RsrS-RsrR two component system was reported in 2016 [9]. Δ*rsrR* and Δ*rsrS* strains were constructed using a marker-less gene knockout method in *A. caldus*. The transcription levels of *rsrS*, *rsrR*, *tetH*, and *doxDA* were analyzed by RT-qPCR under the stimulation of $K_2S_4O_6$ in the wild type and two gene knockout strains, and the results indicated that the RsrS-RsrR regulated the transcription of *tetH* and *doxDA* in a $K_2S_4O_6$–dependent manner. The regulatory protein RsrR was expressed and purified to verify protein and promoter DNA binding using electrophoretic mobility shift assays (EMSAs). A 19 bp (AACACCTGTTACACCTGTT) inverted repeat sequence (IRS) was identified to be the binding motif of RsrR through EMSA and promoter probe plasmid analysis *in vitro* and *in vivo*, respectively. Hence, as summarized in **Figure 3**, the RsrS can sense the extracellular tetrathionate signal and autophosphorylated, then RsrR is activated by receiving the phosphate from RsrS, the phosphorylated RsrR dimerizes and binds on the IRS region of *tetH* operon and initiates the transcription of the genes together with the RNA polymerase.

The RR protein of TCS can function as the activator of $\sigma^{54}$-dependent transcription initiation, which converts the closed RNAP-$\sigma^{54}$ holoenzyme complex to open state to initiate transcription. $\sigma^{54}$ -dependent RR proteins have been reported in several bacteria [21–23]. It was reported that the two-component system TspS-TspR could regulate the sulfur oxidization (Sox) system in *Acidithiobacillus caldus* and some chemolithotrophic bacteria in a $\sigma^{54}$-dependent manner [10]. RT-PCR was used to analyze the

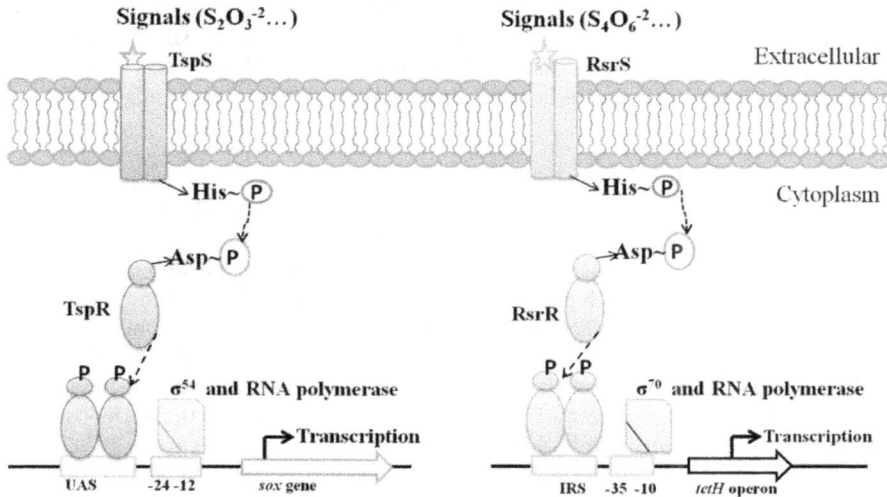

**Figure 3.**
*Different TCS regulation mechanisms between the sox system and the $S_4I$ pathway. Sox system and the $S_4I$ pathway are important sulfur oxidation system in A. caldus, and the regulation mechanism of which has been revealed. The regulation mechanism was summarized in this diagram, the left part represented the regulation mechanism of TspS-TspR on the Sox system in A. caldus summarized according to literature [10]. The right part was the model of RsrS-RsrR regulating the $S_4I$ pathway in A. caldus [9]. The activation signals, the interaction between the regulators and key binding motifs of the promoters were showed.*

composition of the *sox* operon. Results indicated that the genes in the *sox* operon were cotranscribed whereas the transcription of *tspR* was independent. The activation of $\sigma^{54}$ on the transcription of the *sox* genes was verified by the higher transcripts of the operon genes in the constructed *rpoN*-overexpression strain. Following, the transcription initiation site (TSS) of the first gene (*sox*-X) in the operon was verified using 5'RACE (Rapid amplification of cDNA ends). Upstream of the TSS (G, +1), the potential −12 (GC) and − 24 (GG) sites were also identified, which was a typical feature of the $\sigma^{54}$-dependent promoter. Promoter-probe plasmids were constructed to analyze the promoter activity in *A. caldus* by comparing the wild type (P1) and the mutated promoter (GG/GC mutated to AA, $P_{12M}$, $P_{24M}$, and $P_{12/24M}$) containing strains. Hence, the $\sigma^{54}$-dependent promoter was verified by 5'RACE and promoter-probe plasmid activity analysis. As reported, the $\sigma^{54}$-dependent transcription requires binding of the enhancer binding protein (EBP) to upstream activator sequences (UASs) to activate transcription initiation. TspR protein was then expressed and purified to analyze the binding of P1 promoter by EMSA. The binding of TspR protein with different length promoters with two different predicted UASs was analyzed, and only the promoter containing $UAS_1$ (TGTCCCAAATGGGACA) showed a shift lane on the native PAGE. To verify the critical sites in $UAS_1$, $UAS_{M1}$ and $UAS_{M2}$ mutants were constructed, which converted the bases TGT/ACA to GAG/GTG and changed the variable bases CCC to TTG, respectively. TGT/ACA was identified as the critical sites of $UAS_1$ by analyzing the activity of the wild type, $UAS_{M1}$ and $UAS_{M2}$ mutants. Thus, the experiments confirmed that the Sox system was regulated by the $\sigma^{54}$-dependent two-component system TspS/TspR. A signal transduction and transcriptional regulation model for the Sox system in *A. caldus* is depicted in **Figure 3**. TspS can sense the signal stimuli such as thiosulfate and other sulfur substrates, and phosphorylate at a proposed conserved His residue. TspS is activated TspR by the transferring phosphoryl group from its His residue to the conserved Asp residue of TspR. Subsequently, the activated TspR is dimerized and binds to the UAS sequence of promoter $P_1$, meanwhile changing the conformation of the holoenzyme ($\sigma^{54}$-RNA polymerase), which binds to the $-12/-24$ region to activate the transcription of *sox* genes. Interestingly, potential $-12/-24$ region and UAS sequences were predicted in other bacteria with similar *tspS-tspR-sox* gene composition, which may indicate the importance of TCS in the regulation the Sox oxidation system.

## 4. Roles of two-component system in ferrous iron oxidation

*A. ferrooxidans* is an important iron and sulfur oxidizing bacterium in the *Acidithiobacillus* genus, which can oxidize Fe (II) and reduced sulfur compounds to obtain energy for growth. Compared with Fe (II), sulfur seems to be a better energy source because it can provide more ATP at the same molar level [24]. Understanding the function and regulation mechanism of the two energy production systems is critical in coordinating sulfur and iron oxidation process to avoid the $S^0$ deposition and improve the efficiency of bioleaching.

When *A. ferrooxidans* was cultivated in the presence of both Fe (II) and $S^0$ as electron donors, the Fe (II) concentration, bacterial concentration, and pH were measured along with the growth process [7]. The results indicated that ferrous iron was oxidized before $S^0$. The redox potential increased in the Fe (II) oxidization process, while kept stable during sulfur oxidation. Additionally, RT-qPCR analysis showed that the Fe (II) oxidization genes (*rus, cup, petC1* and *cta*) transcribed before the sulfur oxidation genes (*cyoB, hdrA, hdrC, hdrB, sqr* and *tetH*). The sensor/regulator two-component signal transducing system RegBA consisting of a redox-sensing RegB and a DNA binding RegA, located near the *rus* operon, was also studied. The recombinant RegA was produced and purified and was used in the EMSA experiments to analyze its binding with related gene promoters. The retarded bands could be detected with the regulatory regions of the Fe (II) oxidization genes (*rus, petI, cta*) and sulfur oxidation genes (*cyo, hdr, hdrB, tetH* and *sqr*), which indicated the regulation roles of RegA on these genes. As a result, an initial model of the RegBA regulation on *A. ferrooxidans* sulfur and iron oxidation was proposed. Both the full-length RegA and the DNA binding domain of RegA could bind the *rus* and *hdr* operon regulatory regions in the phosphorylated and unphosphorylated state [13]. However, the recombinant RegA tagged with six histidines had signs of aggregation and precipitation. Moinier et al. attempted to purify the RegA protein using the SUMO tag and compared its binding affinity to the target genes in four different states [12]. Similarly, different forms of RegA (DNA binding domain, wild-type, unphosphorylated and phosphorylated-like forms) are able to specifically bind to the regulatory region of the *rus, cta, petI, reg, tet, cyo, hdr,* and *sqr* genes/operons, and the binding of the target genes leads to the formation of multimeric complexes as shown by EMSA results. Further, dynamic light scattering (DLS) analysis also confirmed that 6His-SUMO-RegA protein was polydispersed according to the increased size of hydrodynamic diameters. The protein in the solution (94.6%) had a mean diameter of 7.58 nm, indicating that it was a stable dimer without target binding, whereas it multimerized in the presence of its target DNA, which was consistent with the EMSA analysis. Acetyl phosphate and amino acid mutation was used to change the phosphorylation state of RegA and all the treatments showed that the phosphorylation state of RegA had no effect on the binding affinity for the targets. TSS was determined for the iron oxidation genes (*rus, cta, petI,* and *reg* operons) and RISC oxidation genes (*tetH, cyo, hdr, hdrB, cyd, sqr* and *doxII* operons) using 5′ RACE experiments. The main promoters were $\sigma^{70}$-dependent whereas the *tetH* gene and the *cyd* operons had predicted $\sigma^{54}$-dependent promoters and the *cyd* operon also had a $\sigma^{70}$-dependent promoter. The sequences and the downstream sequences of the RNA polymerase binding site of each gene were amplified and used in the analysis. Results indicated that RegA mainly binded upstream of the $-10$ ($-12$) /$-35$ ($-24$) region, except for the PI promoter of the *rus* operon and the *tetH* operon promoter. However, no RegA binding motifs could be found in the binding gene promoter region using several bioinformatics analysis methods. Hence, the RegBA is a global regulatory system regulating the expression of genes involved in the energy production.

Moreover, other regulatory proteins may be involved in the regulation of these genes. The transcription factor Fur was proven to control the transcription of *petI* operon by binding to the promoter region in EMSA experiments [25]. Fur may inhibit the binding of RNA polymerase and repress the transcription of *petI* in the presence of high intracellular levels of Fe (II). RegA may impede Fur binding on the regulatory region of the *petI* operon when Fe (II) is present and activate its transcription. However, the interaction between RegA and Fur requires further studies. Interestingly, the identification of $\sigma^{54}$-dependent promoters in *tetH* operon was parallel to the occurrence of a $\sigma^{54}$-dependent transcriptional response regulator and cognate histidine kinase at the upstream of *tetH* operon whereas the role of the TCS has not been verified. A $\sigma^{54}$-dependent transcriptional regulator was also predicted at the upstream of the *cyd* operon consistent with the existence of this type promoter whereas no histidine kinase was found near the gene [26].

Based on the reported results, the regulation model for RegBA two-component system is portrayed in **Figure 4**. When Fe (II) is used as the electro donor, RegB is able to sense the low potential state and activate through autophosphorylation. It then activates the RegA protein by transferring the phosphoryl group to the conserved Asp residue of RegA. Phosphorylated RegA protein multimerizes and binds to the promoter region of the target genes, which may activate iron oxidation genes by repressing the binding of other repressor proteins such as Fur for *petI* as well as

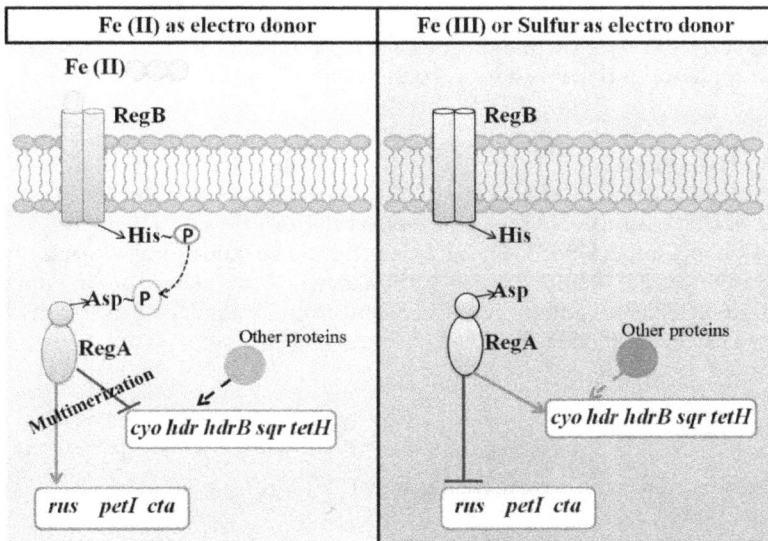

**Figure 4.**
*Regulation of sulfur and ferrous iron oxidation by the TCS system. The regulation in A. ferrooxidans is complicated because it can oxidize Fe (II) and reduced sulfur compounds to obtain energy for growth. Hence, the regulation in this bacterium can be divided according to the energy resources used. The left part represents the regulation mechanism when Fe is the electro donor, and the right part represents the case when sulfur is the electro donor. Several genes are involved in the regulation summarized according to the studies reported [7, 12, 13, 25, 27–29].*

repress sulfur oxidation by interaction with other activator proteins such as $\sigma^{54}$-dependent transcriptional regulator for *tetH*. In the absence of Fe (II), RegB is not activated and other proteins may act together with RegA, leading to the activation of sulfur oxidation genes and repression of the iron oxidation genes. The interaction between RegBA and other regulatory proteins should be studied further to fully understand the regulation mechanism of the iron and sulfur oxidation pathways.

## 5. Conclusions

Two component systems possess critical roles in the regulation of sulfur and iron oxidation in acidophilic bacteria. In the sulfur oxidizing species *A. caldus*, two typical regulatory modes were identified, the $\sigma^{54}$-dependent TCS regulation in the Sox system and the $\sigma^{70}$-dependent TCS regulation in the $S_4I$ pathway. Meanwhile, research on the global regulatory RegBA system indicates that it could control the transcription of several important genes relevant to iron and sulfur oxidation pathways in *A. ferrooxidans*. Although it has been verified that three different two-component systems can participate the regulation of energy production processes in *Acidithiobacillus* spp., further studies are required in the following aspects: (i) the distribution of similar regulatory systems such as TspS-R, RsrS-R, and RegB-A were identified, but the verification of their regulatory roles in relative genes awaits further research; (ii) the detailed regulation mechanism of the different two-component systems in the iron oxidation and sulfur oxidation bacteria merits investigation, for example, the $\sigma^{54}$-dependent TCS in the *tetH* operon in *A. ferrooxidans*; (iii) studies should examine the interactions between the TCS systems and other regulatory proteins to understand the concrete mechanism of energy regulation in *Acidithiobacillus* spp.; (iv) studies should identify the signaling molecules and reveal the interaction between the signals and the response proteins; (v) the structural studies of the TCSs in *Acidithiobacillus* spp. await further research. Therefore, studies on important TCSs in acidophilic bacteria will benefit the understanding of the mechanisms of their environmental adaption and growth as well as facilitate applications that take advantage their special properties.

## Acknowledgements

This work was supported by grants from the National Natural Science Foundation of China (31900116), the Scientific and Technological Projects of Henan Province (202102310395), the Natural Science Foundation of Shandong Province (6622320549) and the Medical Science and Technology Projects of Henan Province (LHGJ20190955).

## Conflict of interests

The authors declare no conflict of interests.

## Author details

Lifeng Li[1*] and Zhaobao Wang[2]

1 Henan Neurodevelopment Engineering Research Center for Children, Henan Key Laboratory of Children's Genetics and Metabolic Diseases, Children's Hospital Affiliated to Zhengzhou University, Henan Children's Hospital, Zhengzhou Children's Hospital, Zhengzhou, China

2 Energy-rich Compounds Production by Photosynthetic Carbon Fixation Research Center, Shandong Key Lab of Applied Mycology, College of Life Sciences, Qingdao Agricultural University, Qingdao, China

*Address all correspondence to: lsbks1017@126.com

## IntechOpen

# References

[1] Kelly, D.P. and A.P. Wood, *Reclassification of some species of Thiobacillus to the newly designated genera Acidithiobacillus gen. nov., Halothiobacillus gen. nov. and Thermithiobacillus gen. nov.* International Journal of Systematic and Evolutionary Microbiology 2000.**50**: p. 511-516.

[2] Wang, R., et al., *Sulfur Oxidation in the Acidophilic Autotrophic Acidithiobacillus spp.* Front Microbiol, 2018.**9**: p. 3290.

[3] Harrison, A.P., Jr., *The acidophilic thiobacilli and other acidophilic bacteria that share their habitat.* Annu Rev Microbiol, 1984. **38**: p. 265-92.

[4] Zhang, S., et al., *Acidithiobacillus ferrooxidans and its potential application.* Extremophiles, 2018. **22**(4): p. 563-579.

[5] Yang, L., et al., *Acidithiobacillus thiooxidans and its potential application.* Appl Microbiol Biotechnol, 2019. **103**(19): p. 7819-7833.

[6] Dopson, M. and E.B. Lindstrom, *Potential role of thiobacillus caldus in arsenopyrite bioleaching.* Appl Environ Microbiol, 1999. **65**(1): p. 36-40.

[7] Ponce, J.S., et al., *Acidithiobacillus ferrooxidans oxidizes ferrous iron before sulfur likely through transcriptional regulation by the global redox responding RegBA signal transducing system.* Hydrometallurgy, 2012. **127-128**: p. 187-194.

[8] Gomez-Mejia, A., G. Gamez, and S. Hammerschmidt, *Streptococcus pneumoniae two-component regulatory systems: The interplay of the pneumococcus with its environment.* International Journal of Medical Microbiology, 2018. **308**(6): p. 722-737.

[9] Wang, Z.B., et al., *The Two-Component System RsrS-RsrR Regulates the Tetrathionate Intermediate Pathway for Thiosulfate Oxidation in Acidithiobacillus caldus.* Front Microbiol, 2016. **7**: p. 1755.

[10] Li, L.-F., et al., *The σ54-dependent two-component system regulating sulfur oxidization (Sox) system in Acidithiobacillus caldus and some chemolithotrophic bacteria.* 2017. **101**(5): p. 2079-2092.

[11] Rzhepishevska, O.I., et al., *Regulation of a novel Acidithiobacillus caldus gene cluster involved in metabolism of reduced inorganic sulfur compounds.* Appl Environ Microbiol, 2007. **73**(22): p. 7367-72.

[12] Moinier, D., et al., *The Global Redox Responding RegB/RegA Signal Transduction System Regulates the Genes Involved in Ferrous Iron and Inorganic Sulfur Compound Oxidation of the Acidophilic Acidithiobacillus ferrooxidans.* Front Microbiol, 2017. **8**: p. 1277.

[13] Moinier, D., et al., *How the RegBA Redox Responding System Controls Iron and Sulfur Oxidation in Acidithiobacillus ferrooxidans.* Advanced Materials Research, 2013. **825**: p. 186-189.

[14] Merrick, M.J., *In a class of its own-- the RNA polymerase sigma factor sigma 54 (sigma N).* Mol Microbiol, 1993. **10**(5): p. 903-9.

[15] Buck, M., et al., *The bacterial enhancer-dependent sigma(54) (sigma(N)) transcription factor.* J Bacteriol, 2000. **182**(15): p. 4129-36.

[16] Reitzer, L. and B.L. Schneider, *Metabolic context and possible physiological themes of sigma(54)-dependent genes in Escherichia coli.* Microbiol Mol Biol Rev, 2001. **65**(3): p. 422-44, table of contents.

[17] Bordes, P., et al., *The ATP hydrolyzing transcription activator*

*phage shock protein F of Escherichia coli: identifying a surface that binds sigma 54.* Proc Natl Acad Sci U S A, 2003. **100**(5): p. 2278-83.

[18] Lardi, M., et al., *σ54-Dependent Response to Nitrogen Limitation and Virulence in Burkholderia cenocepacia Strain H111.* Appl Environ Microbiol, 2015. **81**(12): p. 4077-89.

[19] Kazakov, A.E., et al., *σ54-dependent regulome in Desulfovibrio vulgaris Hildenborough.* BMC Genomics, 2015. **16**: p. 919.

[20] Bush, M. and R. Dixon, *The role of bacterial enhancer binding proteins as specialized activators of σ54-dependent transcription.* Microbiol Mol Biol Rev, 2012. **76**(3): p. 497-529.

[21] Kern, D., et al., *Structure of a transiently phosphorylated switch in bacterial signal transduction.* Nature, 1999. **402**(6764): p. 894-8.

[22] Park, S., et al., *Two-component signaling in the AAA + ATPase DctD: binding Mg2+ and BeF3- selects between alternate dimeric states of the receiver domain.* Faseb j, 2002. **16**(14): p. 1964-6.

[23] Sallai, L. and P.A. Tucker, *Crystal structure of the central and C-terminal domain of the sigma(54)-activator ZraR.* J Struct Biol, 2005. **151**(2): p. 160-70.

[24] Findlay, A.J. and A. Kamyshny, *Turnover Rates of Intermediate Sulfur Species ([Formula: see text], S(0), S(2) [Formula: see text], S(4)[Formula: see text], [Formula: see text]) in Anoxic Freshwater and Sediments.* Front Microbiol, 2017. **8**: p. 2551.

[25] Lefimil, C., et al., *Regulation of Expression of the PetI Operon Involved in Iron Oxidation in the Biomining Bacterium Acidithiobacillus Ferrooxidans.* Advanced Materials Research, 2009. **71-73**: p. 199-202.

[26] Zhang, Y., et al., *Complete Genome Sequence of Acidithiobacillus Ferrooxidans YNTRS-40, a Strain of the Ferrous Iron- and Sulfur-Oxidizing Acidophile.* Microorganisms, 2019. **8**(1).

[27] Quatrini, R., et al., *Extending the models for iron and sulfur oxidation in the extreme acidophile Acidithiobacillus ferrooxidans.* BMC Genomics, 2009. **10**: p. 394.

[28] Levican, G., et al., *Characterization of the petI and res operons of Acidithiobacillus ferrooxidans.* J Bacteriol, 2002. **184**(5): p. 1498-501.

[29] Elsen, S., et al., *RegB/RegA, a highly conserved redox-responding global two-component regulatory system.* Microbiol Mol Biol Rev, 2004. **68**(2): p. 263-79.

Chapter 3

# Thriving at Low pH: Adaptation Mechanisms of Acidophiles

*Xianke Chen*

## Abstract

Acid resistance of acidophiles is the result of long-term co-evolution and natural selection of acidophiles and their natural habitats, and formed a relatively optimal acid-resistance network in acidophiles. The acid tolerance network of acidophiles could be classified into active and passive mechanisms. The active mechanisms mainly include the proton efflux and consumption systems, generation of reversed transmembrane electrical potential, and adjustment of cell membrane composition; the passive mechanisms mainly include the DNA and protein repair systems, chemotaxis and cell motility, and quorum sensing system. The maintenance of pH homeostasis is a cell-wide physiological process that adopt differently adjustment strategies, deployment modules, and integration network depending on the cell's own potential and its habitat environments. However, acidophiles exhibit obvious strategies and modules similarities on acid resistance because of the long-term evolution. Therefore, a comprehensive understanding of acid tolerance network of acidophiles would be helpful for the intelligent manufacturing and industrial application of acidophiles.

**Keywords:** acidophiles, acid-resistance, pH homeostasis, adaptation, evolution

## 1. Introduction

Both natural and man-made acidic habitats are widely distributed in global land and ocean ecosystems, such as acidic sulfur-rich thermal springs, marine volcanic vents, and acid mine drainage (AMD) [1]. However, these unique habitats harbor the active acidophilic organisms that are well adapted to the acidic environments. Undoubtedly, acidophiles are distributed randomly throughout the tree of life and prevalent in the acidity or extreme acidity habitats, archaea and bacteria in particular, and they represent an extreme life-forms [2–4]. Generally, acidophilic archaea and bacteria mainly include members of phylum *Euryarchaeota*, *Crenarchaeota*, *Proteobacteria*, *Acidobacteria*, *Nitrospira*, *Firmicutes*, *Actinobacteria* and *Aquificae* such as *Ferroplasma*, *Acidiplasma*, *Sulfolobus*, *Acidianus*, *Acidiphilum*, *Acidithiobacillus*, *Acidihalobacter*, *Ferrovum*, *Acidiferrobacter*, *Acidobacterium*, *Leptospirillum*, *Sulfobacillus*, *Acidibacillus*, *Acidimicrobium*, and *Hydrogenobaculum* [5–7]. More importantly, acidophiles, as an important taxa of microorganisms, are closely related to the biogeochemistry cycles, eco-environment and human development, such as driving the elemental sulfur and iron cycles [8], the water and soil polluted by acidic effluents [9], biomining-bioleaching techniques and bioremediation technologies [9–11]. Thus, a comprehensive understanding of the acid-resistance networks and modules of acidophiles would be helpful for the

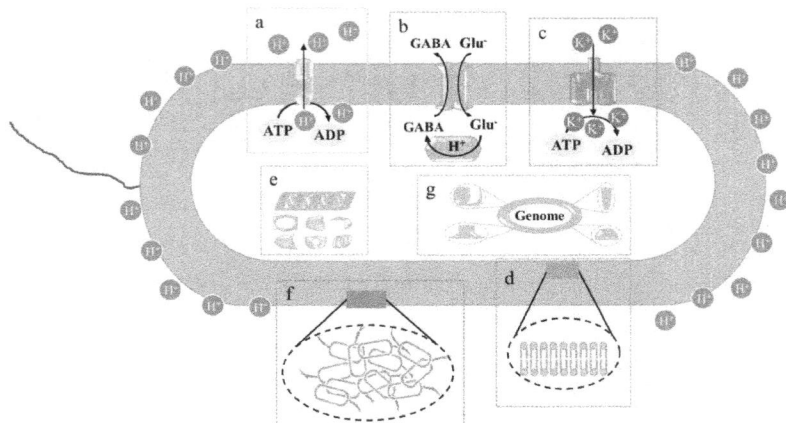

**Figure 1.**
*The active and passive acid-resistance mechanisms in acidophiles. (a) Proton pump: F₁F₀–ATPase complex pump protons out of the cells though ATP hydrolysis. (b) Proton consumption modules: GadB-GadC system can consume excess intracellular protons. (c) Reversed transmembrane electrical potential (Δψ) modules: Generating a reversed Δψ is by positive ions transport (e.g. K⁺ transport). (d) Membranes system: The highly impermeable cell membranes structure. (e) Macromolecules protection modules: A larger proportion of DNA and protein repair systems such as Dps, GrpE, MolR, and DnaK proteins. (f) Escaping system: QS system, biofilm, chemotaxis and cell motility modules. (g) Other modules: Some possible mechanisms of imperfect classification, including iron "rivet", degradation proteins of organic acids, surface proteins of high pI values, and outer membrane porin.*

understanding of the evolutionary processes, ecological behaviors and industry applications of acidophiles.

Acidophiles thrive at an extremely low pH and maintain a relatively neutral cytoplasm pH [12], namely maintenance several orders of magnitude difference in proton concentrations in cell; thus, one of the main challenges to these microorganisms living in acidic habitats is the extremely acidic stress environments. Acidophiles have evolved a large number of mechanisms to withstand the deleterious effects of fluctuations in proton concentration (**Figure 1**), due to the fact that acidophiles face the challenge of maintaining a near neutral intracellular pH. Currently, the mechanisms of growth and acid tolerance of typical extreme acidophiles in extremely low pH environments have been widely studied [13–15]. Herein, we, specifically focusing on acid-tolerant mechanisms, strategies, functions, and modules instead of species types, reviewed and summarized the current knowledge of the acid-resistance networks adopted by acidophiles for coping with acid or extreme acid environments. In addition, owing to space constraints and complexity of acidophiles types, we limit our discussion of the acid-tolerant adaptation mechanisms to typical acidophiles (archaea and bacteria) that populate acidic habitats.

## 2. Acid-resistant mechanisms of acidophiles

### 2.1 Active support of acidophiles pH homeostasis

Microorganisms tend to maintain a high proton motive force (PMF) and a near-neutral pH in cytoplasm. The transmembrane electrical potential (Δψ) and transmembrane pH gradient (ΔpH) could vary as a function of the external pH. The immediately available energy source for acidophilic cell is this pre-existing transmembrane proton gradient, due to the external environments are frequently in the pH range of 1.0–3.0, while the typical pH of cytoplasms are close to 6.5

(that is, the differential proton concentrations of 4–6 orders of magnitude). The ΔpH across the membrane is a major part of the PMF, and the ΔpH is linked to cellular bioenergetics. Acidophiles, such as *Acidithiobacillus ferrooxidans* and *Acidithiobacillus caldus*, are capable of using the ΔpH to generate a large quantity of ATP [16, 17]. However, this processes would lead to the rapid acidification of the cytoplasm of alive cells. Because a high level of protons concentration would destroy essential molecules in cell, such as DNA and protein, acidophiles have evolved the capability to pump protons out of their cells at a relatively high rate. The $F_1F_o$–ATPase consists of a hydrophilic part ($F_1$) composed of α, β, γ, δ, and ε subunits and a hydrophobic membrane channel ($F_o$) composed of a, b, and c subunits; among them, the $F_1$ catalyzes ATP hydrolysis or synthesis and the $F_o$ translocates protons. This mechanism pumps out protons from cells by hydrolyzing ATP (**Figure 1**), thereby efficiently protecting cells from the acidic environments. In several microorganisms, transcriptional level of the *atp* operon upregulated by exposure to the acidic environments, including *A. caldus*, *Acidithiobacillus thiooxidans*, and *Lactobacillus acidophilus* [18–20], suggesting its critical role in acid resistance of cell. Several proton efflux proteins have also been identified in the sequenced genomes of *A. ferrooxidans*, *A. thiooxidans*, *A. caldus*, *Ferroplasma acidarmanus*, and *Leptospirillum* group II [21, 22]. The $H^+$-ATPase activity and $NAD^+$/NADH ratio were upregulated in *A. thiooxidans* under the acid stress [19]. The cells actively pump out protons by a respiratory chain from cell. For example, under the acid stress, the *A. caldus* increases its expression of respiratory chain complexes that can pump protons out of the cells [20]. Meanwhile, $NAD^+$ involved in glycolysis as the coenzyme of dehydrogenase, generating large amount of ATP and contributing to pump protons out of the cells though ATP hydrolysis.

Among the active mechanisms, the proton consumption systems are necessary to remove excess intracellular protons. Once protons enter the cytoplasm, some mechanisms and patterns are required to mitigate effects caused by a high concentration of proton in cells. Under the acidic conditions, there is increased expression of amino acid decarboxylases enzymes (such as Glutamate decarboxylase-β (GadB)) that could consume the cytoplasmic protons by the catalytic reactions [23]. GadB, coupling with a glutamate/gamma-aminobutyrate antiporter (GadC), catalyzed glutamate to γ-aminobutyric acid (GABA) and exchanged with glutamate substrate to achieve continued decarboxylation reactions (**Figure 1**) [24]. It consumed a proton during the decarboxylation reactions and thus supported the intracellular pH homeostasis. And, it would contribute to a reversed Δψ in most bacteria. Similarly, the *gadB* gene was found in *Ferroplasma* spp., and the gene transcription was upregulated under acid shock conditions in *A. caldus* [20, 22]. Therefore, in order to maintain pH homeostasis of cell, acidophiles need to be able to consume excess protons from the cytoplasm.

A second major strategy for the active mechanisms used by acidophiles to reduce the influx of protons is the generation of an inside positive Δψ that generated by a Donnan potential of positively charged ions. A positive inside transmembrane potential was contributed to a reversed Δψ that could prevent protons leakage into the cells. The acidophiles might use the same strategies to generate a reversed membrane potential to resist the inward flow of protons, $Na^+$/$K^+$ transporters in particular (**Figure 1**) [25]. Previous data showed that some genomes of acidophiles (*A. thiooxidans*, *F. acidarmanus*, *Sulfolobus solfataricus*, etc.) contain a high number of cation transporters genes and these transporters were probably involved in the generation of Donnan potential to inhibit the protons influx [21, 22, 25, 26]. The genome of *Picrophilus torridus* also encodes large number of proton-driven secondary transporters which represents adaptation to the more extremely acidic environment [27]. Furthermore, we found that the maintenance of Δψ in *A. thiooxidans* was directly related to the uptake of cations, especially the influx of potassium ions [25]. Further evidence of chemiosmotic gradient created by a Donnan potential to support acid resistance is the Donnan potential created by a passive mechanism, that is, a small residual inside positive $\Delta\psi$ and $\Delta pH$ are maintained in inactive cells of *A. caldus*, *A. ferrooxidans*, *Acidiphilium acidophilum*, and *Thermoplasma acidophilum* [28–30]. The residual $\Delta\psi$ and $\Delta pH$ studies have been criticized because of measurement methods [31]. However, subsequent data showed that the energy-dependent cation pumps played an important role in generating an inside positive $\Delta\psi$. In addition, acidophilic bacteria are highly tolerant to cations and are more sensitive to anions. In summary, the inside positive $\Delta\psi$ is a ubiquitous and significant strategy in maintaining the cellular pH homeostasis.

Although improving the efflux and consumption of protons and increasing the expression of secondary transporters are a common strategy, the most effective strategy is also to reduce the proton permeability of cell membrane [32, 33]. Acidophiles can synthesize a highly impermeable membrane to respond to proton attack (**Figure 1**). These physiological adaptations membranes are composed of the high levels of iso/anteiso-BCFAs (branched chain fatty acids), both saturated and mono-unsaturated fatty acids, $\beta$-hydroxy, $\omega$-cyclohexyl and cyclopropane fatty acids (CFAs) [34]. It was found that cell membrane resisted the acid stress by increasing the proportion of unsaturated fatty acid and CFAs in some bacteria, such as *A. ferrooxidans* and *Escherichia coli* [35–37]. Although the cytoplasmic membrane is the main barrier to protons influx, the destruction of the membrane caused by protons may cause this barrier to break down. The key component of membranes preventing acid damage seems to be CFAs, which contributes to the formation of cell membrane compactness. Supporting this mechanism is that *E. coli* with a mutation in the *cfa* gene became quite sensitive to low pH and can overcome this sensitivity by providing the exogenous *cfa* gene [36]. Meanwhile, the transcription of *cfa* gene was upregulated under the acid stress in *A. caldus* [20], and it suggests that changing the fatty acid content of the cell membrane is an adaptive response to acid stress. In brief, the CFAs is important for maintaining membrane integrity and compactness under the acid conditions.

To maintain the pH homeostasis of cells, acidophilic archaea cells have a highly impermeable cell membrane to restrict proton influx into the cytoplasm. One of the key characteristics of acidophilic archaea is the monolayer membrane typically composed of large amount of GDGTs, which are extremely impermeable to protons [38–40]. Although acidophilic bacteria have a variety of acid-resistant adaptation strategies, compared with acidophilic archaea, it has not been found that these bacteria would exhibit excellent growth ability below pH 1. The special tetraether lipid is closely related to acid-tolerance capability, because the ether linkages are less sensitive to acid hydrolysis than ester linkages [41]. And, the results of studies on acidophilic archaea indicated that tetraether lipids may be more resistant to acid than previously thought [42]. Therefore, the contribution of tetraether lipids to adaptation of archaea to extremely low pH is enormous. To a certain extent, it also supports the reason why dominance of archaea under extremely acidic environments. Similarly, the extreme acid tolerance of archaea can be attributed to cyclopentane rings and the vast methyl-branches [43]. In addition, it was found that the less phosphorus in the lipoprotein layer of acidophilus cell can contribute to higher hydrophobicity, which was beneficial for resisting extreme acid shock [13]. Irrespective of the basic composition of cell membranes, bacteria and archaea have extensively reshaped their membrane components to overcome the extremely low acid environments. In summary, the impermeable of acidophilic cell membrane is an important strategy for the pH homeostasis of acidophiles formed by restricting the influx of protons into the cells.

## 3. Passive strategies for acidophiles living

When the cells are attacked or stressed by higher concentrations of protons, the passive mechanisms of pH homeostasis would support the active mechanism. If protons penetrate the acidophilic cell membrane, a range of intracellular repair systems would help to repair the damage of macromolecules [13]. The DNA and protein repair systems play a central role in coping with acid stress of cells (**Figure 1**). Because DNA carries genetic information of cell life and protein plays an important role in the physiological activities of cells, DNA or protein damage caused by protons would bring irreversible harm to cells. When the cells are exposed to a high concentration of proton environments or protons influx into the cells, a great number of DNA repair proteins and chaperones (such as Dps, GrpE, MolR, and DnaK protein) would repair the damaged DNA and protein [19, 44, 45]. Previously reported study showed that a great number of DNA and protein repaired genes presence in wide range of extreme acidophiles genomes might be related to the acid resistance, for example, a large number of the DNA repaired proteins genes in *P. torridus* genome [27, 46]. Indeed, the transcription and expression of these repair systems were upregulated under the extreme acid stress, for example, the transcription of molecular chaperones repair system-molR and DnaK were enhanced in *A. thiooxidans* [19]. In addition, the GrpE and DnaK proteins expression were significantly improved in *Acetobacter pasteurianus* for coping with acetic acid stress [47]. Similarly, the molecular chaperones involved in protein refolding were largely expressed in *L. ferriphilum* under the AMD biofilm communities [48]. And, the chaperones were also highly expressed in *F. acidarmanus* during aerobic culture [49].

Quorum sensing (QS) system is a ubiquitous phenomenon that establishes the cell to cell communication in a population through the production, secretion and detection of signal molecules. In addition, The QS system is also widely involved in various physiological processes in cell such as biofilm formation, exopolysaccharides, motility, and bacterial virulence [50–52]. Moreover, the QS system can contribute to bacteria tolerating extreme environmental conditions by regulated biofilm formation. For example, bacteria showed the strong resistance to extremely low pH, due to these bacteria grown in a biofilm environment [53]. In case of acidophiles, QS system has been reported in *A. ferrooxidans* by producing the stable acylated homoserine lactones (AHLs) signal molecules under acidic conditions and overexpression strains promoted cell growth by regulated genes expression [54, 55].

Flagella is an important cell structure for the motility and chemotaxis in most bacteria, and is also involved in the biofilm formation [56]. Flagella-mediated chemotaxis is essential for cells to respond to environmental stimuli (pH, temperature, osmotic pressure, etc.) and find nutrients for growth. The chemotaxis and motility of cells is a complex physiological behavior regulated by the diverse transcription factors, such as RpoF ($\sigma^{28}$ or FliA) of the σ factors and ferric uptake regulator (Fur) of the global regulator, and has strictly spatiotemporal characteristics [20, 56]. For example, the mutant strain of *A. caldus fur* gene significantly upregulated some genes (*cheY*, *cheV*, *flhF*, *flhA*, *fliP*, *fliG*, etc.) related to cell chemotaxis and motility under the acid shock conditions [20]. Similarly, *F. acidarmanus* was capable of motility and biofilm formation [57]. This indicates that although the chemotaxis and cell motility ability of acidophiles cannot directly involve in acid resistance and maintain cell pH homeostasis, they have the ability to avoid extremely unfavorable acid environments to improve cells survival. Altogether, we suggest that the QS system and chemotaxis and cell motility are essential part of escaping the extremely acidic environments in passive mechanisms (**Figure 1**).

It could be seen from the classification description above that there are a variety of mechanisms and strategies by which acidophiles can tolerate or resist the acidic or extremely acidic environments. However, some possible mechanisms have been imperfectly understood or classified, for example, the distinctive structural and functional characteristics of extremely acidophilic microorganisms (**Figure 1**) [13, 15]. First, iron may act as a "rivet" at low pH, which plays an important role in maintaining proteins activity, for example, the high proportion of iron proteins in *F. acidiphilum*. And, it has been found that the removal of iron from proteins can result in the loss of proteins activity [58, 59]. Secondly, the strategy of cell surface charges. The surface proteins of acidophiles have a high pI values (a positive surface charges), which can act as a transient proton repellent on the cell surface. For example, the isoelectric point (pI) of the OmpA-like protein in the outer membrane of the *A. ferrooxidans* is 9.4, whereas that of *E. coli* OmpA is 6.2 [60]. It may be the functional requirements that the possession of positive surface charges could reduce the permeability of *A. ferrooxidans* cells to protons. Then, adjustment of pore size of membrane channels is also used to minimize inward proton leakage under acid stress. For example, under the acid shock, the expression of outer membrane porin (Omp40) of *A. ferrooxidans* was upregulated [61], which could control the size and ion selectivity of the entrance to the pore. Ultimately, since organic acids could diffuse into the cells in the form of protonation at low pH environments and then the proton dissociation quickly acidify the cytoplasm, the degradation of organic acids might be a potential mechanism for maintaining pH homeostasis, especially heterotrophic acidophiles. Although the genes that degrade organic acids in some acidophile (such as *F. acidarmanus, P. torridus*) have been identified, it is unclear whether the degradation of organic acids would contribute pH homeostasis [27, 62]. In summary, these possible mechanisms remain to be confirmed but these genes of existence and identification could be a mechanism associated with low pH tolerance.

## 4. Evolution of low pH fitness of acidophiles

In the past few decades, studies have confirmed that acidophilic microorganisms are widely present in the three domains of bacteria, archaea and eukarya, indicating that acidophiles have gradually developed in the evolution of life on earth, rather than from a single adaptation events. Although the extremely acidic environments are toxic to most organisms, there are still large number of indigenous microorganisms that can thrive in these habitats. The generally accepted view is that acidophiles can be divided into moderate acidophiles that have pH optima of between 3 and 5, extreme acidophiles that have pH optima for growth at pH < 3, and hyperacidophiles that have pH optima for growth below pH 1 [1]. Generally, with the acidity becomes more extreme, biodiversity also gradually decreases. Accordingly, as would be anticipated, the most extremely acidic environments hold the less biodiversity, for example, hyperacidophiles includes the relatively few species (e.g. *F. acidarmanus* and *Picrophilus oshimae*) [1]. Acidophiles can survive in the acidic or extremely environments and are the source of acidity environment [1, 63, 64]; thus, they have the ability to resist the acidic environments that evolved during evolution.

Acidic hydrothermal ecosystems, such as Tengchong hot springs, Crater Lake, and Yellowstone National Park, are dominated by archaea [40, 65], and suggesting that the acidophilic archaea evolved in the extremely acidic hydrothermal environments after the emergence of oxygenic photosynthesis [66]. Based on the niche similarity and physiological adaptation among archaea, it showed that the long-term acidity stress is the main selection pressure to control the evolution of archaea and leads to the co-evolution of acid-resistant modules [66]. Although the species diversity decreases significantly as the pH decreases, the high abundance of acidophilic taxa, such as *Gammaproteobacteria* and *Nitrospira*, was detected in acid habitats. Indeed, for the dominant lineages such as *Acidithiobacillus* spp. and *Leptospirillum* spp., this pH-specific niche partitioning was obvious [67]. Consistent with this, *Ferrovum* is more acid-sensitive than *A. ferrooxidans* and *L. ferrooxidans*, and prefers to grow under the near-moderate pH [68]. Interestingly, the majority of acidophiles growing at extremely acidic (i.e. pH < 1) are heterotrophic acidophiles that are capable of utilizing organic matter for growth such as *T. acidophilum* and *P. torridus*. In addition, although the *Acidiplasma* spp. and *Ferroplasma* spp. can oxidize ferrous iron in biomining, organic carbon can also be used for growth, and their relative abundance would increase with the mortality of other bioleaching microorganisms [69, 70]. Therefore, they can be regarded as scavengers of the dead microorganisms and help the material and energy cycle in acidic habitats. To sum up, coexisting species may occupy different niches that could be affected by the pH changes, resulting in the changes in their distribution patterns.

The reasons for the dominance of these particular microorganisms in acidic ecosystems are presumed to their adaptive capabilities. Adaptations to acid stress dictate the ecology and evolution of the acidophiles. Acidic ecosystems are a unique ecological niche for acid-adapted microorganisms. These relatively low-complexity ecosystems offer a special opportunity for the evolutionary processes and ecological behaviors analyses of acidophilic microorganisms. In the last decade, the use of high-throughput sequencing technology and post-genomic methods have significantly promoted our understanding of microbial diversity and evolution in acidic environments [68]. At present, metagenomics studies have revealed various acidophilic microorganisms from environments such as the AMD and acidic geothermal areas, and found that these microorganisms play an important role in acid generation and adaptability to the environments [71, 72]. For example, because the comparative metagenomics and metatranscriptomics directly recover and reveal microbial genome information from the environments, it has the potential to provide insights into acid-resistance mechanisms of the uncultivated bacteria, such as *clpX*, *clpP*, and *sqhC* genes for resistance against acid stress. In addition, metatranscriptomics and metaproteomics analyses further uncovered the major metabolic and adaptive capabilities in situ [71], indicating the mechanisms of response and adaptation to the extremely acid environments.

The continuous exploration of acidic habitats and acidophilic microorganisms is the basis for comprehending the evolution of acidophilic microbial acid-tolerant modules, strategies, and networks. First, methods based on transcriptomics and proteomics are the key to understanding the global acid-tolerant network of individuals under acid stress [19, 73]. Secondly, comparative genomics plays a vital role in exploring the acid adaptation mechanism of acidophiles and studying the evolution of acidophiles genomes [74]. Ultimately, the emerging metagenomics technologies play an important role in evaluating and predicting microbial communities and their adaptability to acidic environments [75]. Moreover, metagenomics approaches could also provide a large amount of knowledge and functional module analysis on the acid tolerance of acidophiles to fully develop their potential in the evolution of acid tolerance [76]. With the publication of large number of metagenomics data, the evolution of the acid-tolerant components of these extremophiles would be better illustrated in the future.

## 5. Conclusions

Understanding the maintenance of pH homeostasis in acidophiles is of great significance to comprehend the mechanisms of cells growth and survival, as well as to the eco-remediation and application of biotechnology; thus, it is essential to fully understand the acid-tolerant networks and strategies of acidophilic microorganisms. The aims of this chapter presents the acid-resistant modules and strategies of acidophiles in more detail, including the proton efflux and consumption, reversed membrane potential, impermeable cell membrane, DNA and protein repair systems, and QS system (**Figure 1**). However, at present, several of the pH homeostatic mechanisms still lack clear and rigorous experimental evidence to support their functions from my point of view. In addition, we also discussed the evolution of acidophiles and its acid-resistant modules. In brief, the true purpose of acidophilic microorganisms evolving these mechanisms is to tolerate the extremely acidic environments or reduce its harmful effects for cell survival.

Acidophiles are known for their remarkable acid resistance. Over the last decades, the combination of molecular and biochemical analysis of acidophiles with genome, transcriptome, and proteome have provided new insights into the acid-resistant mechanisms and evolution of the individual acidophiles at present. Using these genome sequences in a functional context through the application of high throughput transcriptomic and proteomic tools to scrutinize acid stress might elucidate further potential pH homeostasis mechanisms. However, the disadvantages of genomics, transcriptomics, and proteomics are that the data are descriptive and analogous and more work is required to verify the hypotheses such as the mutational analyses and genetic markers. One of the main obstacles to the current research on acid tolerance of acidophiles is the lack of genetic tools for in-depth analysis. Therefore, the development of genetic tools and biochemical methods in acidophile would facilitate elucidating the molecular mechanisms of acidophile adapting to extremely acidic environments, such as vector development remain largely unexplored. In addition, as most acidophiles are difficult to isolate and culture, our ability to understand acid resistance of acidophile is limited. The emerging omics technologies would be a crucial step to explore the spatiotemporal transformation patterns of acidophilic microbial communities, microbial ecophysiology and evolution in the future.

## Acknowledgements

We are grateful to Pro. Jianqiang Lin from Shandong University for providing the opportunity to review the acid-resistant mechanisms of acidophiles and Pro. Linxu Chen (Shandong University) for suggestions and comments on the outline and manuscript. We also thank Wenwen Xiang (Xiamen University) and Yujie Liu (Shandong University) for English language editing.

## Conflict of interest

We declare no conflicts of interest.

## Author details

Xianke Chen[1,2]

1 Research Center for Eco-Environmental Sciences, Chinese Academy of Sciences, Beijing, China

2 Sino-Danish College of University of Chinese Academy of Sciences, Beijing, China

*Address all correspondence to: chenxianke20@mails.ucas.ac.cn

## IntechOpen

# References

[1] Johnson DB, Aguilera A. Extremophiles and Acidic Environments. In: Schmidt TM, editor. Encyclopedia of Microbiology (Fourth Edition). Oxford: Academic Press; 2019. p. 206-27. DOI:10.1016/B978-0-12-809633-8.90687-3

[2] Meyer-Dombard DA, Shock E, Amend J. Archaeal and bacterial communities in geochemically diverse hot springs of Yellowstone National Park, USA. Geobiology. 2005;3:211-27. DOI:10.1111/j.1472-4669.2005.00052.x

[3] Mirete S, Morgante V, González-Pastor JE. Acidophiles: Diversity and Mechanisms of Adaptation to Acidic Environments. In: Stan-Lotter H, Fendrihan S, editors. Adaption of Microbial Life to Environmental Extremes: Novel Research Results and Application. Cham: Springer International Publishing; 2017. p. 227-51. DOI:10.1007/978-3-319-48327-6_9

[4] Arce-Rodríguez A, Puente-Sánchez F, Avendaño R, Martínez-Cruz M, de Moor JM, Pieper DH, et al. Thermoplasmatales and sulfur-oxidizing bacteria dominate the microbial community at the surface water of a CO2-rich hydrothermal spring located in Tenorio Volcano National Park, Costa Rica. Extremophiles. 2019;23(2):177-87. DOI:10.1007/s00792-018-01072-6

[5] Dopson M. Physiological and Phylogenetic Diversity of Acidophilic Bacteria. 2016. p. 79-92. DOI:10.21775/9781910190333.05

[6] Golyshina O, Ferrer M, Golyshin P. Diversity and Physiologies of Acidophilic Archaea. 2016. p. 93-106. DOI:10.21775/9781910190333.06

[7] Chen L-x, Huang L-n, Méndez-García C, Kuang J-l, Hua Z-s, Liu J, et al. Microbial communities, processes and functions in acid mine drainage ecosystems. Current Opinion in Biotechnology. 2016;38:150-8. DOI:10.1016/j.copbio.2016.01.013

[8] Druschel GK, Baker BJ, Gihring TM, Banfield JF. Acid mine drainage biogeochemistry at Iron Mountain, California. Geochemical Transactions. 2004;5(2):13. DOI:10.1186/1467-4866-5-13

[9] Johnson DB, Hallberg KB. Acid mine drainage remediation options: a review. Science of The Total Environment. 2005;338(1):3-14. DOI:10.1016/j.scitotenv.2004.09.002

[10] Clark ME, Batty JD, van Buuren CB, Dew DW, Eamon MA. Biotechnology in minerals processing: Technological breakthroughs creating value. Hydrometallurgy. 2006;83(1):3-9. DOI:10.1016/j.hydromet.2006.03.046

[11] Hedrich S, Johnson DB. Remediation and Selective Recovery of Metals from Acidic Mine Waters Using Novel Modular Bioreactors. Environmental Science & Technology. 2014;48(20):12206-12. DOI:10.1021/es5030367

[12] Matin A. Keeping a neutral cytoplasm; the bioenergetics of obligate acidophiles. FEMS Microbiology Letters. 1990;75(2):307-18. DOI:10.1016/0378-1097(90)90541-W

[13] Baker-Austin C, Dopson M. Life in acid: pH homeostasis in acidophiles. Trends in Microbiology. 2007;15(4):165-71. DOI:10.1016/j.tim.2007.02.005

[14] Slonczewski JL, Fujisawa M, Dopson M, Krulwich TA. Cytoplasmic pH Measurement and Homeostasis in Bacteria and Archaea. In: Poole RK, editor. Advances in Microbial Physiology. 55: Academic

Press; 2009. p. 1-317. DOI:10.1016/S0065-2911(09)05501-5

[15] Krulwich TA, Sachs G, Padan E. Molecular aspects of bacterial pH sensing and homeostasis. Nature Reviews Microbiology. 2011;9(5):330-43. DOI:10.1038/nrmicro2549

[16] Dopson M, Lindström BE, Hallberg KB. ATP generation during reduced inorganic sulfur compound oxidation by Acidithiobacillus caldus is exclusively due to electron transport phosphorylation. Extremophiles. 2002;6(2):123-9. DOI:10.1007/s007920100231

[17] Apel WA, Dugan PR, Tuttle JH. Adenosine 5'-triphosphate formation in Thiobacillus ferrooxidans vesicles by H+ ion gradients comparable to those of environmental conditions. Journal of bacteriology. 1980;142(1):295-301. DOI:10.1128/JB.142.1.295-301.1980

[18] Kullen MJ, Klaenhammer TR. Identification of the pH-inducible, proton-translocating F1F0-ATPase (atpBEFHAGDC) operon of Lactobacillus acidophilus by differential display: gene structure, cloning and characterization. Molecular Microbiology. 1999;33(6):1152-61. DOI:10.1046/j.1365-2958.1999.01557.x

[19] Feng S, Yang H, Wang W. System-level understanding of the potential acid-tolerance components of Acidithiobacillus thiooxidans ZJJN-3 under extreme acid stress. Extremophiles. 2015;19(5):1029-39. DOI:10.1007/s00792-015-0780-z

[20] Chen X-k, Li X-y, Ha Y-f, Lin J-q, Liu X-m, Pang X, et al. Ferric Uptake Regulator Provides a New Strategy for Acidophile Adaptation to Acidic Ecosystems. Applied and Environmental Microbiology. 2020;86(11):e00268-20. DOI:10.1128/AEM.00268-20

[21] Tyson GW, Chapman J, Hugenholtz P, Allen EE, Ram RJ, Richardson PM, et al. Community structure and metabolism through reconstruction of microbial genomes from the environment. Nature. 2004;428(6978):37-43. DOI:10.1038/nature02340

[22] Ni G, Simone D, Pamla D, Broman E, Wu X, Turner S, et al. A Novel Inorganic Sulfur Compound Metabolizing Ferroplasma-Like Population Is Suggested to Mediate Extracellular Electron Transfer. Front Microbiol. 2018;9. DOI:10.3389/fmicb.2018.02945

[23] Richard H, Foster JW. Escherichia coli Glutamate- and Arginine-Dependent Acid Resistance Systems Increase Internal pH and Reverse Transmembrane Potential. Journal of Bacteriology. 2004;186(18):6032. DOI:10.1128/JB.186.18.6032-6041.2004

[24] Ma D, Lu P, Yan C, Fan C, Yin P, Wang J, et al. Structure and mechanism of a glutamate–GABA antiporter. Nature. 2012;483(7391):632-6. DOI:10.1038/nature10917

[25] Suzuki I, Lee D, Mackay B, Harahuc L, Oh JK. Effect of Various Ions, pH, and Osmotic Pressure on Oxidation of Elemental Sulfur by Thiobacillus thiooxidans. Applied and Environmental Microbiology. 1999;65(11):5163. DOI:10.1128/AEM.65.11.5163-5168.1999

[26] She Q, Singh RK, Confalonieri F, Zivanovic Y, Allard G, Awayez MJ, et al. The complete genome of the crenarchaeon Sulfolobus solfataricus P2. Proceedings of the National Academy of Sciences. 2001;98(14):7835. DOI:10.1073/pnas.141222098

[27] Fütterer O, Angelov A, Liesegang H, Gottschalk G, Schleper C, Schepers B, et al. Genome sequence of Picrophilus torridus and its implications for life around pH 0. Proceedings of

the National Academy of Sciences. 2004;101(24):9091. DOI:10.1073/pnas.0401356101

[28] Cox JC, Nicholls DG, Ingledew WJ. Transmembrane electrical potential and transmembrane pH gradient in the acidophile Thiobacillus ferro-oxidans. Biochemical Journal. 1979;178(1):195-200. DOI:10.1042/bj1780195

[29] Goulbourne Jr E, Matin M, Zychlinsky E, Matin A. Mechanism of ΔpH maintenance in active and inactive cells of an obligately acidophilic bacterium. Journal of Bacteriology. 1986;166(1):59-65. DOI:10.1128/jb.166.1.59-65.1986

[30] Hsung JC, Haug A. Membrane potential ofThermoplasma acidophila. FEBS Letters. 1977;73(1):47-50. DOI:10.1016/0014-5793(77)80012-4

[31] Michels M, Bakker EP. Generation of a large, protonophore-sensitive proton motive force and pH difference in the acidophilic bacteria Thermoplasma acidophilum and Bacillus acidocaldarius. Journal of Bacteriology. 1985;161(1):231-7. DOI:10.1128/jb.161.1.231-237.1985

[32] Siliakus MF, van der Oost J, Kengen SWM. Adaptations of archaeal and bacterial membranes to variations in temperature, pH and pressure. Extremophiles. 2017;21(4):651-70. DOI:10.1007/s00792-017-0939-x

[33] Konings WN, Albers S-V, Koning S, Driessen AJM. The cell membrane plays a crucial role in survival of bacteria and archaea in extreme environments. Antonie van Leeuwenhoek. 2002;81(1):61-72. DOI:10.1023/A:1020573408652

[34] Oger PM, Cario A. Adaptation of the membrane in Archaea. Biophysical Chemistry. 2013;183:42-56. DOI:10.1016/j.bpc.2013.06.020

[35] Mykytczuk NCS, Trevors JT, Ferroni GD, Leduc LG. Cytoplasmic membrane fluidity and fatty acid composition of Acidithiobacillus ferrooxidans in response to pH stress. Extremophiles. 2010;14(5):427-41. DOI:10.1007/s00792-010-0319-2

[36] Chang Y-Y, Cronan JE. Membrane cyclopropane fatty acid content is a major factor in acid resistance of *Escherichia coli*. Molecular Microbiology. 1999;33(2):249-59. DOI:10.1046/j.1365-2958.1999.01456.x

[37] Shabala L, Ross T. Cyclopropane fatty acids improve *Escherichia coli* survival in acidified minimal media by reducing membrane permeability to H+ and enhanced ability to extrude H+. Research in Microbiology. 2008;159(6):458-61. DOI:10.1016/j.resmic.2008.04.011

[38] Jacquemet A, Barbeau J, Lemiègre L, Benvegnu T. Archaeal tetraether bipolar lipids: Structures, functions and applications. Biochimie. 2009;91(6):711-7. DOI:10.1016/j.biochi.2009.01.006

[39] Koga Y, Morii H. Recent Advances in Structural Research on Ether Lipids from Archaea Including Comparative and Physiological Aspects. Bioscience, Biotechnology, and Biochemistry. 2005;69(11):2019-34. DOI:10.1271/bbb.69.2019

[40] Xie W, Zhang CL, Wang J, Chen Y, Zhu Y, de la Torre JR, et al. Distribution of ether lipids and composition of the archaeal community in terrestrial geothermal springs: impact of environmental variables. Environmental Microbiology. 2015;17(5):1600-14. DOI:https://doi.org/10.1111/1462-2920.12595

[41] Mathai JC, Sprott GD, Zeidel ML. Molecular Mechanisms of Water and Solute Transport across Archaebacterial Lipid Membranes. Journal of Biological

Chemistry. 2001;276(29):27266-71. DOI:10.1074/jbc.M103265200

[42] Macalady JL, Vestling MM, Baumler D, Boekelheide N, Kaspar CW, Banfield JF. Tetraether-linked membrane monolayers in Ferroplasma spp: a key to survival in acid. Extremophiles. 2004;8(5):411-9. DOI:10.1007/s00792-004-0404-5

[43] Langworthy TA. Long-chain diglycerol tetraethers from Thermoplasma acidophilum. Biochimica et Biophysica Acta (BBA) - Lipids and Lipid Metabolism. 1977;487(1):37-50. DOI:10.1016/0005-2760(77)90042-X

[44] Tomoyasu T, Ogura T, Tatsuta T, Bukau B. Levels of DnaK and DnaJ provide tight control of heat shock gene expression and protein repair in *Escherichia coli*. Molecular Microbiology. 1998;30(3):567-81. DOI:10.1046/j.1365-2958.1998.01090.x

[45] Calhoun LN, Kwon YM. Structure, function and regulation of the DNA-binding protein Dps and its role in acid and oxidative stress resistance in *Escherichia coli*: a review. Journal of Applied Microbiology. 2010.

[46] Ciaramella M, Napoli A, Rossi M. Another extreme genome: how to live at pH 0. Trends in Microbiology. 2005;13(2):49-51. DOI:10.1016/j.tim.2004.12.001

[47] Andrés-Barrao C, Saad MM, Chappuis M-L, Boffa M, Perret X, Ortega Pérez R, et al. Proteome analysis of Acetobacter pasteurianus during acetic acid fermentation. Journal of Proteomics. 2012;75(6):1701-17. DOI:10.1016/j.jprot.2011.11.027

[48] Ram R, Verberkmoes N, Thelen M, Tyson G, Baker B, Blake R, et al. Community Proteomics of a Natural Microbial Biofilm. Science. 2005;308:1915-20. DOI:10.1126/science. 1109070

[49] Dopson M, Baker-Austin C, Bond PL. Analysis of differential protein expression during growth states of Ferroplasma strains and insights into electron transport for iron oxidation. Microbiology. 2005;151(12):4127-37. DOI:10.1099/mic.0.28362-0

[50] Montgomery K, Charlesworth JC, LeBard R, Visscher PT, Burns BP. Quorum sensing in extreme environments. Life (Basel). 2013;3(1):131-48. DOI:10.3390/life3010131

[51] Abbamondi GR, Kambourova M, Poli A, Finore I, Nicolaus B. Chapter 4 - Quorum Sensing in Extremophiles. In: Tommonaro G, editor. Quorum Sensing: Academic Press; 2019. p. 97-123. DOI:10.1016/B978-0-12-814905-8.00004-6

[52] Mamani S, Moinier D, Denis Y, Soulère L, Queneau Y, Talla E, et al. Insights into the Quorum Sensing Regulon of the Acidophilic Acidithiobacillus ferrooxidans Revealed by Transcriptomic in the Presence of an Acyl Homoserine Lactone Superagonist Analog. Front Microbiol. 2016;7:1365-. DOI:10.3389/fmicb.2016.01365

[53] Li Y-H, Hanna MN, Svensäter G, Ellen RP, Cvitkovitch DG. Cell Density Modulates Acid Adaptation in Streptococcus mutans Implications for Survival in Biofilms. Journal of Bacteriology. 2001;183(23):6875. DOI:10.1128/JB.183.23.6875-6884.2001

[54] Farah C, Vera M, Morin D, Haras D, Jerez CA, Guiliani N. Evidence for a Functional Quorum-Sensing Type AI-1 System in the Extremophilic Bacterium Acidithiobacillus ferrooxidans. Applied and Environmental Microbiology. 2005;71(11):7033. DOI:10.1128/AEM.71.11.7033-7040.2005

[55] Gao XY, Fu CA, Hao LK, Gu XF, Wang R, Lin J, et al. The substrate-dependent regulatory effects of the

AfeI/R system in Acidithiobacillus ferrooxidans reveals the novel regulation strategy of quorum sensing in acidophiles. Environmental Microbiology. 2020. DOI:10.1111/1462-2920.15163

[56] Yang C-L, Chen X-K, Wang R, Lin J-Q, Liu X-M, Pang X, et al. Essential Role of σ Factor RpoF in Flagellar Biosynthesis and Flagella-Mediated Motility of Acidithiobacillus caldus. Front Microbiol. 2019;10:1130-. DOI:10.3389/fmicb.2019.01130

[57] Baker-Austin C, Potrykus J, Wexler M, Bond PL, Dopson M. Biofilm development in the extremely acidophilic archaeon 'Ferroplasma acidarmanus' Fer1. Extremophiles. 2010;14(6):485-91. DOI:10.1007/s00792-010-0328-1

[58] Golyshina OV, Golyshin PN, Timmis KN, Ferrer M. The 'pH optimum anomaly' of intracellular enzymes of Ferroplasma acidiphilum. Environmental Microbiology. 2006;8(3):416-25. DOI:10.1111/j.1462-2920.2005.00907.x

[59] Ferrer M, Golyshina OV, Beloqui A, Golyshin PN, Timmis KN. The cellular machinery of Ferroplasma acidiphilum is iron-protein-dominated. Nature. 2007;445(7123):91-4. DOI:10.1038/nature05362

[60] Chi A, Valenzuela L, Beard S, Mackey AJ, Shabanowitz J, Hunt DF, et al. Periplasmic Proteins of the Extremophile Acidithiobacillus ferrooxidans: A High Throughput Proteomics Analysis*. Molecular & Cellular Proteomics. 2007;6(12):2239-51. DOI:10.1074/mcp.M700042-MCP200

[61] Guiliani N, Jerez CA. Molecular Cloning, Sequencing, and Expression of omp-40 the Gene Coding for the Major Outer Membrane Protein from the Acidophilic Bacterium

Thiobacillus ferrooxidans. Applied and Environmental Microbiology. 2000;66(6):2318. DOI:10.1128/AEM.66.6.2318-2324.2000

[62] Angelov A, Liebl W. Insights into extreme thermoacidophily based on genome analysis of Picrophilus torridus and other thermoacidophilic archaea. Journal of Biotechnology. 2006;126(1):3-10. DOI:10.1016/j.jbiotec.2006.02.017

[63] Johnson D, Quatrini R. Acidophile Microbiology in Space and Time. Current Issues in Molecular Biology. 2020;39:63-76. DOI:10.21775/cimb.039.063

[64] Kirk Nordstrom D, Blaine McCleskey R, Ball JW. Sulfur geochemistry of hydrothermal waters in Yellowstone National Park: IV Acid–sulfate waters. Applied Geochemistry. 2009;24(2):191-207. DOI:10.1016/j.apgeochem.2008.11.019

[65] Ward L, Taylor MW, Power JF, Scott BJ, McDonald IR, Stott MB. Microbial community dynamics in Inferno Crater Lake, a thermally fluctuating geothermal spring. The ISME Journal. 2017;11(5):1158-67. DOI:10.1038/ismej.2016.193

[66] Colman DR, Poudel S, Hamilton TL, Havig JR, Selensky MJ, Shock EL, et al. Geobiological feedbacks and the evolution of thermoacidophiles. The ISME Journal. 2018;12(1):225-36. DOI:10.1038/ismej.2017.162

[67] Kuang J, Huang L, He Z, Chen L, Hua Z, Jia P, et al. Predicting taxonomic and functional structure of microbial communities in acid mine drainage. The ISME Journal. 2016;10(6):1527-39. DOI:10.1038/ismej.2015.201

[68] Kuang J-L, Huang L-N, Chen L-X, Hua Z-S, Li S-J, Hu M, et al. Contemporary environmental variation determines microbial diversity

patterns in acid mine drainage. The ISME Journal. 2013;7(5):1038-50. DOI:10.1038/ismej.2012.139

[69] Golyshina OV, Timmis KN. Ferroplasma and relatives, recently discovered cell wall-lacking archaea making a living in extremely acid, heavy metal-rich environments. Environmental Microbiology. 2005;7(9):1277-88. DOI:10.1111/j.1462-2920.2005.00861.x

[70] Golyshina OV, Tran H, Reva ON, Lemak S, Yakunin AF, Goesmann A, et al. Metabolic and evolutionary patterns in the extremely acidophilic archaeon Ferroplasma acidiphilum YT. Scientific Reports. 2017;7(1):3682. DOI:10.1038/s41598-017-03904-5

[71] Chen L-x, Hu M, Huang L-n, Hua Z-s, Kuang J-l, Li S-j, et al. Comparative metagenomic and metatranscriptomic analyses of microbial communities in acid mine drainage. The ISME Journal. 2015;9(7):1579-92. DOI:10.1038/ismej.2014.245

[72] Menzel P, Gudbergsdóttir SR, Rike AG, Lin L, Zhang Q, Contursi P, et al. Comparative Metagenomics of Eight Geographically Remote Terrestrial Hot Springs. Microbial Ecology. 2015;70(2):411-24. DOI:10.1007/s00248-015-0576-9

[73] Mangold S, rao Jonna V, Dopson M. Response of Acidithiobacillus caldus toward suboptimal pH conditions. Extremophiles. 2013;17(4):689-96. DOI:10.1007/s00792-013-0553-5

[74] Zhang X, Liu X, He Q, Dong W, Zhang X, Fan F, et al. Gene Turnover Contributes to the Evolutionary Adaptation of Acidithiobacillus caldus: Insights from Comparative Genomics. Front Microbiol. 2016;7:1960-. DOI:10.3389/fmicb.2016.01960

[75] Cowan DA, Ramond JB, Makhalanyane TP, De Maayer P.

Metagenomics of extreme environments. Current Opinion in Microbiology. 2015;25:97-102. DOI:10.1016/j.mib.2015.05.005

[76] Cárdenas JP, Quatrini R, Holmes DS. Genomic and metagenomic challenges and opportunities for bioleaching: a mini-review. Research in Microbiology. 2016;167(7):529-38. DOI:10.1016/j.resmic.2016.06.007

Chapter 4

# Quorum Sensing of Acidophiles: A Communication System in Microorganisms

*Xueyan Gao, Jianqiang Lin, Linxu Chen, Jianqun Lin and Xin Pang*

## Abstract

Communication is important for organisms living in nature. Quorum sensing system (QS) are intercellular communication systems that promote the sociality of microbes. Microorganisms could promote cell-to-cell cooperation and population density to adapt to the changing environment through QS-mediated regulation that is dependent on the secretion and the detection of signal molecules (or called autoinducers). QS system is also discovered in acidophiles, a microorganism that is widely used in the bioleaching industry and can live in an acidic environment. An example is the LuxI/R-like QS system (AfeI/R) that has been reported in the chemoautotrophic species of the genus *Acidithiobacillus*. In this chapter, we will introduce the types and distribution of the QS system, and the biological function and regulatory mechanism of QS in acidophiles. We will also discuss the potential ecological function of QS system and the application value of the QS system in the control and regulation of the bioleaching process in the related industries and acid mine damage.

**Keywords:** quorum sensing, communication, signal molecules, environmental adaption

## 1. Introduction

Acidophiles is a microorganism that can live in an acidic environment and widely distributed in extremely harsh environments such as acid mines, sulfur-containing hot springs, and volcanic craters [1, 2]. The signal communication and cooperation of the bacterial flora could be conducive to the survival and propagation of acidophiles in an extremely harsh environment. Quorum sensing (QS), as an important part of sociomicrobiology, is a group behavior that enables bacteria to establish cell-to-cell communication by producing, secreting, and detecting signal molecules (also called autoinducers) [3–5]. With the increase of cell density, the concentration of signal molecules released by cells becomes higher. When the concentration of signal molecules accumulates to a threshold in the local environment, the signal molecules bind to the receptor protein to activate or inhibit the expression of specific genes and then allow bacteria to respond to population density and external environment [6, 7]. Diverse biological functions of bacteria are regulated by QS systems, such as the formation of biofilm, the production of antibiotics, the expression of pathogenic virulence genes, the luminescence of marine organisms, the transfer of Ti plasmids, and so on [8–11].

The research of the QS system has a history of 50 to 60 years [12]. As early as 1965, Tomasz and Alexander reported the interesting phenomenon caused by QS [13]. Hormone-like cell products could control the competence of *Pneumococcus* and promote foreign DNA to enter the cell [13]. Subsequently, Nealson et al. found that the luminescence of marine bacteria was positively correlated with the quorum of bacteria [14]. The research of the QS system had a major breakthrough with the identification of the signal molecules of the QS system and the discovery of *luxI/R* operon in *Vibrio ficheri* and the regulatory mechanism of the QS system regulating the luminescence phenotype of *V. ficheri* was revealed [15–17]. 3-O-C$_6$-HSL was confirmed as the signal molecule of the QS system of *V. ficheri*, which was synthesized by the synthetase encoded by *luxI* gene. In contrast, the *luxR* gene encodes the transcription factor that binds to the signal molecules. LuxI/R-mediated QS system regulates the expression of the *lux* operon, thereby affecting the luminescence phenotype of *V. ficheri* [15–17]. In recent years, more and more signal molecules, regulatory mechanisms, and functions of QS system have been discovered, with the help of modern analytical experimental techniques such as bioinformatics, molecular biology, and chemical analysis.

Compared with the well-studied QS systems in model bacteria and some pathogenic bacteria, the studies on QS system in acidophiles are restricted due to the limitations in molecular manipulation techniques. In the present chapter, we will introduce the research on the QS system of acidophiles, outline the distribution and molecular mechanisms of the QS system in acidophiles, and discuss the function of the QS system involved in the control and regulation of the bioleaching process in the biomining industry and acid mine damage.

## 2. Quorum sensing of acidophiles

*Acidithiobacillus* spp. is an important member of the acidophilic and chemo-lithotrophic Gram-negative sulfur-oxidizing bacteria [18–20]. Members of *Acidithiobacillus* can grow on reduced inorganic sulfur compounds (RISCs), and some of them can use ferrous as the energy substrate [18–20]. In 2005, Farah et al. discovered the LuxI/R-type QS system (AfeI/R) in *Acidithiobacillus ferrooxidans* [21]. The AfeI/R system is composed of three genes located on one operon (*afeI-orf3-afeR*) [21, 22]. The AfeI is a homologous protein to LuxI, catalyzing the synthesis of signal molecules, while the AfeR is homologous to LuxR protein, functioning by recognizing and binding signal molecules and regulating gene expression. The *Orf3* locus, located between the *AfeI* and *AfeR*, has a mysterious existence and its function is presently unknown [21, 22].

Bioinformatics analysis revealed that the AfeI/R-type QS system is widely distributed in the nine species of *Acidithiobacillus* reported so far (**Figure 1**) [23]. AfeI/R-like QS system could be identified from *Acidithiobacillus ferrooxidans*, *Acidithiobacillus ferridurans*, and *Acidithiobacillus ferrivorans*, which means that in addition to *Acidithiobacillus ferrianus* and *Acidithiobacillus ferriphilus*, the AfeI/R system was found in almost all the sulfur- and ferrous iron-oxidizing species of *Acidithiobacillus*. However, among the sulfur-oxidizing species of *Acidithiobacillus*, only *Acidithiobacillus thiooxidans* was reported to have a QS system. It is worth noting that the conserved *afeR-orf3-afeI*-type operon can be found in *A. ferrooxidans*, *A. ferridurans*, and *A. thiooxidans*, while in *A. ferrivorans* were separated *afeI* and *afeR* genes. Besides, lower homology proteins were found in *A. thiooxidans* ATCC19377.

AfeI/R-like QS system also found in the genus of sulfur-oxidizing-only bacterium *Thiomonas* and the *Acidiferrobacter* reportedly produce acyl-HSL [24]. Further

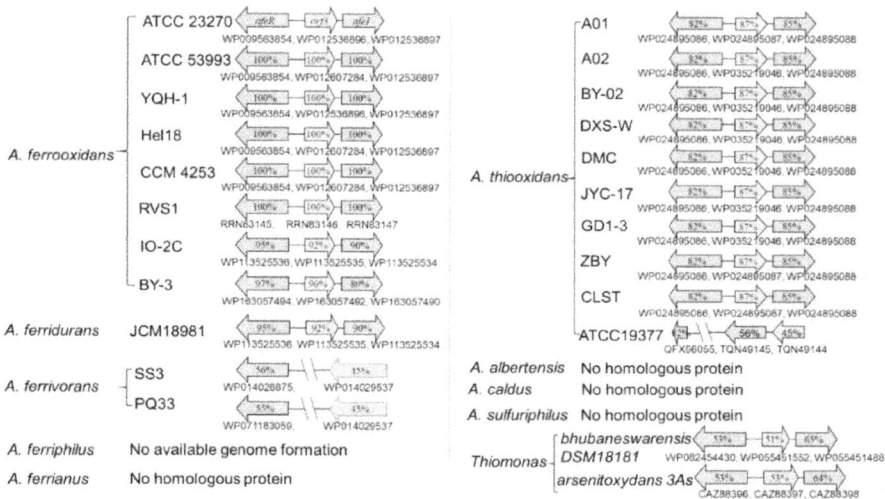

**Figure 1.**
*Distribution of AfeI/R-like QS system in Acidithiobacillus and other acidophiles [23].*

analysis of the acidophilus strains revealed that AfeI homologous proteins were found in 7 genera of acidophiles, and AfeR homologous proteins were present in 17 genera, and Orf3 homologous proteins were found in 12 genera of acidophiles (**Table 1**). AfeI/R-like QS system showed some variations at the gene arrangement and protein sequence in acidophilus strains.

| AfeI homologous | AfeR homologous | Orf3 homologous |
|---|---|---|
| Acidithiobacillus | Acidithiobacillus | Acidithiobacillus |
| Acidocella | Acidianus | Acidianus |
| Frateuria | Acidilobus | Acidomonas |
| Metallosphaera | Acidimicrobium | Aciduliprofundum |
| Nitrosotalea | Acidiphilium | Caldivigra |
| Sulfurisphaera | Acidisphaera | Ferrimicrobium |
| Thiomonas | Acidocella | Ferrovum |
| | Acidomonas | Nitrosotalea |
| | Caldisphaera | Picrophilus |
| | Caldivigra | Sulfolobus |
| | Desulfosporosinus | Thiobacillus |
| | Ferrovum | Thiomonas |
| | Frateuria | |
| | Leptospirillum | |
| | Picrophilus | |
| | Thiobacillus | |
| | Thiomonas | |

**Table 1.**
*Distribution of AfeI/R- like QS system in acidophiles based on the AfeI/R sequence of Acidithiobacillus ferrooxidans alignment in acidophiles.*

In 2007, another QS system (QS-II) of *A. ferrooxidans* was discovered [25]. This QS system includes four co-transcribed genes, *glyQ*, *glyS*, *gph*, and *act*, which encode α and β subunits of glycine t-RNA synthetase, phosphatase, and acyltransferase [25]. The reporter bacteria and GC–MS technology confirmed that acyl-HSL ($C_{14}$-HSL) could be synthesized by the *act* gene heterologously expressed in *Escherichia coli* [25]. Semi-quantitative RT-PCR experiments showed that the expression of *act* gene was higher when cultured in $Fe^{2+}$-enriched media than that in $S^0$-enriched media [25]. However, the regulatory protein that corresponds to Act has not been discovered, and whether the Act can synthesize acyl-HSLs or other types of signal molecules in *A. ferrooxidans* has not been reported up to now. Therefore, whether the Act system is a functional QS system is still controversial [26]. There are still many unsolved mysteries of *act* operon, which are worthy of in-depth study in the future.

It is worth noting that a thermophilic-like ene-reductase (*foye*-1) was discovered in the acidophilic iron-oxidizing bacterium *Ferrovum* sp. [27]. The *foye-1* gene is located directly upstream of *luxR* gene, and the encoded protein played a major role in intercellular communication [7]. Since the *luxI* gene is absent on the genome of *Ferrovum sp*. JA12, it is speculated that FOYE-1 may replace LuxI and participate in the QS process [27].

Interestingly, a diffusible signal factor (DSF) quorum sensing system was deciphered in the acidophilic, ferrous-oxidizing species, *Leptospirillum ferriphilum* [28]. This kind of QS system consisted of the *rpf* operon, which contains four genes, *rpfF-rpfC-rpfC-rpfG*, of which the *rpfF* gene encodes DSF synthase, the *rpfC* genes encode Hpt domain-containing protein and signal transduction kinase, and the *rpfG* gene encodes a two-component system response regulator [28]. Besides, three LuxR family transcriptional regulator proteins and another autoinducer-binding domain-containing protein were reported in *L. ferriphilum* [28]. Sequence alignment found the possible DSF perception protein *rpfR* and the homologous protein of *rpfC* in *A. caldus* and *Sulfobacillus thermosulfidooxidans* [29]. The homologous protein of *rpfG* was also found in *S. thermosulfidooxidans* [29].

Therefore, previous research results and bioinformatics analysis indicated that the QS system is universal and unique in acidophiles. Some of the acidophiles such as *A. ferrooxidans*, *A. ferridurans, and A. thiooxidans* have the complete AfeI/R-type QS system, which can synthesize and respond to signal molecules. This type of QS system is considered to be a fully functional and undisputed QS system. Some of the acidophiles only contain AfeI homologous protein that has the function of synthesizing signal molecules. Some acidophiles exist with the orphan LuxR family protein and the DSF-type QS system. Besides, more than one QS system were reported in some acidophiles.

## 3. Types of signal molecules synthesized by the QS system

There are many types of signal molecules synthesized and secreted by the QS system. The N-acyl homoserine lactone (acyl-HSL) is the prominent and widely studied signal molecule of the QS system and is composed of a homoserine lactone ring and an amide side chain (**Figure 2**) [4, 7]. The functional group of the third carbon atom has three forms: hydrogen, hydroxyl, and carbonyl [4, 7]. The R chain group can be 4–18 carbons, with or without an unsaturated C-C bond [12]. The terminal carbon has a branch in some bacteria, and the R chain group in some bacteria is an aromatic acid [12]. Furanosyl borate ester was reported to be the signal molecules used by the AI-2-type QS system [4]. Quinolone, diffusible signaling factor (DSF), hydroxyl-palmitic acid methyl ester (PAME), and small peptide have been reported as signal molecules for the QS system [4, 12].

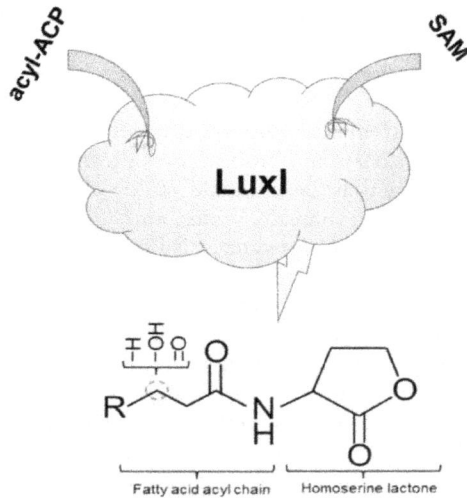

**Figure 2.**
*Structure and synthesis process of acyl-HSL.*

In 2005, Farah et al. reported that nine acyl-HSLs, namely 3-OH-$C_8$-HSL, 3-OH-$C_{10}$-HSL, $C_{12}$-HSL, 3-OH-$C_{12}$-HSL, 3-O-$C_{12}$-HSL, $C_{14}$-HSL, 3-OH-$C_{14}$-HSL, 3-O-$C_{14}$-HSL, and 3-OH-$C_{16}$-HSL, were detected in *A. ferrooxidans* ATCC 23270 cultured in ferrous, elemental sulfur, and thiosulfate energy [21]. The type and function of signal molecules produced by the AfeI/R system are confused due to the potential Act-type QS system in *A. ferrooxidans* [25]. In 2020, Gao et al. determined the types of signal molecules produced by AfeI by the construction of the gene mutant strain of *afeI* and *act* [23]. Different types and concentrations of acyl-HSLs were synthesized in $S^0$- and $Fe^{2+}$-enriched media of *A. ferrooxidans*, while the acyl-HSLs displayed different functions under different energy conditions. The homoserine lactone of acyl-HSLs is derived from S-adenosylmethionine (SAM), whereas its acyl side chain is derived from fatty acid metabolism and is provided by acyl carrier protein (acyl-ACP) (**Figure 2**) [30–32]. The difference in the molecular structure of acyl-HSLs depends on the length of the acyl side chain provided by the acyl carrier protein and the substituent on the third carbon atom [30–32]. The differences in the metabolism of sulfur and ferrous iron under different culture conditions may cause different kinds of acyl side chains produced in *A. ferrooxidans*, which in turn affects the type of signal molecules synthesized by AfeI [21, 31, 33]. Therefore, the synthesis process of acyl-HSLs by *A. ferrooxidans* and other acidophilic sulfide and iron-oxidizing bacteria under different energy sources will also be the key work of future research, which will help to clarify the close relationship between the AfeI/R system and energy substrates.

It has been reported in many papers that the phenotype of acidophilus bacteria such as *A. ferrooxidans*, *A. ferrivorans*, *L. ferrooxidans*, and *Acidiferrobacter sp.* strain SPIII/3 was affected by acyl-HSLs addition and these acyl-HSLs that influenced the growth and metabolism of acidophiles were not synthesized by the strain itself [24, 26, 29, 34, 35]. Besides, the synthetic tetrazole analog of acyl-HSLs could also stimulate the differential gene expression in *A. ferrooxidans* [36]. Therefore, it can be inferred that cross-communication exists in acidophilus bacteria and related research work needs to be carried out in-depth. In addition to the classic acyl-HSLs-type signal molecules, the DSF was also described in *L. ferriphilum* [28]. Whether there are other types of signal molecules in acidophilus bacteria remains to be studied to enrich the types of signal molecules of the QS system in acidophiles.

## 4. The regulatory function of the QS system in acidophiles

### 4.1 QS system and biofilm formation

The quorum sensing system is an important way to regulate extracellular polymeric substance (EPS) synthesis and biofilm formation [37–39]. Transcriptome data of *A. ferrooxidans* show that the tetrazole analog of acyl-HSLs stimulated the differential expression of more than 100 genes, and 42.5% of the differentially expressed genes are related to biofilm synthesis [36]. Laser confocal microscopy and atomic force microscopy observed that the addition of acyl-HSLs could affect the formation of biofilm formed by acidophiles on the surface of the pyrites [23, 24, 26]. It is worth noting that the influence of signal molecules on the formation of biofilms was specific to the types of acyl-HSLs and bacterial species [14, 24]. Studies have shown that the addition of $C_{14}$-HSL promoted the aggregation of *A. ferrooxidans* and *A. ferrivorans* cells, and enhanced the formation of biofilms on the surface of pyrite [24]. However, although the addition of C12-HSL increased the biofilm formation of *A. ferrooxidans*, it inhibited the biofilm formation of *A. ferrivorans* under the same conditions [24]. Gao et al. found that overexpression of *afeI* could not only promote the EPS synthesis and biofilm formation, but also increased the sulfur oxidation ability of cells and enhanced the erosion effect of *A. ferrooxidans* cells on sulfur [23]. In *A. thiooxidans*, studies confirmed that the QS system positively regulated the *pel* operon, participated in regulating the exopolysaccharide biosynthesis, and then affected the formation of biofilms [35]. The *pelD* gene is located in the *pel* operon and encodes the c-di-GMP-binding protein [35]. The interaction between QS system and c-di-GMP pathway to regulated EPS synthesis and biofilm formation had been displayed in other bacterial species [40–42]. It is reported that the QS system regulates the expression of *pelD* gene and may also regulate some gene-encoding proteins with c-di-GMP synthase activity and/or c-di-GMP degradation activity, resulting in the change in intracellular c-di-GMP levels and in turn affecting the formation of biofilm in *A. thiooxidans* [35].

The regulation of the QS system on the dispersion of biofilms has been confirmed in many bacteria such as *Xanthomonas campestris*, *Pseudomonas aeruginosa*, *and Staphylococcus aureus* [43]. The phenomenon of biofilm dispersal is also found in acidophilus bacteria. The addition of DSF family signal molecules caused biofilm dispersal of *L. ferriphilum* and *S. thermosulfidooxidans*, which strongly inhibited the growth and metabolism of bioleaching bacteria [29].

### 4.2 The regulatory function of AfeI/R in different energy substrate environments of *A. ferrooxidans*

Compared with the QS system in other acidophiles, the research of QS system in *A. ferrooxidans* is more in-depth. As early as 2005, semi-quantitative RT-PCR measurements found that the expression levels of *afeI* and *afeR* genes in $S^0$-enriched media were higher than those in $Fe^{2+}$-enriched media [21], which suggested that the AfeI/R system may function optimally in an environment or medium containing sulfur. Confocal laser microscopy and atomic force microscopy techniques have observed that the AfeI/R-mediated QS system can enhance the formation of biofilms of *A. ferrooxidans* on elemental sulfur or metal sulfide surfaces [26]. The *lux-box* (LuxR family protein-binding sequence) in *A. ferrooxidans* was predicted *via* a bioinformatic approach [44], and the *lux-box* sequence upstream of the *afeI* gene was confirmed *via in vitro* experiments [36]. Due to the difficulty of gene manipulation of *A. ferrooxidans* and other acidophiles, it was not until 2020 that

Gao et al. successfully knocked out and overexpressed the *afeI* gene to determine the regulatory role of AfeI/R in the sulfur and iron culture system [23].

Gao et al. revealed that AfeI/R not only played an important role in $S^0$-enriched media, but also had a more significant regulatory role in $Fe^{2+}$-enriched media [23]. In $S^0$-media, overexpression of *afeI* could increase the cell concentration and acid production capacity of the strain in the lag phase and exponential phase, but did not affect the final population density of the culture system, Therefore, AfeI/R can be used as an "accelerator" for *A. ferrooxidans* cultured in $S^0$-enriched media. Besides, the effect of *afeI* overexpression on EPS synthesis was consistent with that on the sulfate yield and cell density of *A. ferrooxidans* in $S^0$-enriched media. Therefore, the regulation of AfeI/R on EPS synthesis was the key strategy for AfeI/R to regulate cell growth, metabolism, and population density of *A. ferrooxidans*. Moreover, the AfeI/R-regulating EPS synthesis could also be an important way for *A. ferrooxidans* to adapt to the sulfur substrate in the environments. In $Fe^{2+}$-enriched media, overexpression of *afeI* significantly inhibited the cell concentration and the ferrous oxidation capacity of the strain. Therefore, AfeI/R can be used as an "inhibitor," regulating cell metabolic growth and the final population density in $Fe^{2+}$-enriched media. The overexpression of *afeI* significantly inhibited the expression of the hydrogenase synthesis gene cluster (AFE_0700–0719) in *A. ferrooxidans* in $Fe^{2+}$-enriched media, suggesting that AfeI/R may regulate hydrogen metabolism to influence the growth, metabolism, and quorum of the *A. ferrooxidans* cells during ferrous culture. The results indicated that the QS system may have a new regulation function when *A. ferrooxidans* is cultivated with ferrous iron, and research on the related molecular mechanism is needed. Energy substrates can affect the acyl-HSLs synthesized by AfeI and the regulatory effects of AfeI/R in *A. ferrooxidans*, and the substrate-dependent regulation strategy of the AfeI/ R was proposed based on these research findings.

## 5. The application of QS in bioleaching

The bioleaching bacteria, as an important class of acidophiles, are widely distributed in the acid mine environments [2]. The progress of mineral dissolution and metal leaching requires the consortium of the bioleaching bacteria and the attachment of cells to the surface of the ores [45, 46]. The QS system regulates cell aggregation and adsorption, EPS synthesis, and biofilm formation; thus, the QS-mediated regulation could be involved in the regulation of the bioleaching process. Therefore, the QS system has important application value in the bioleaching industry and the treatment of acid pollution.

In 2013, González et al. found that the addition of C12/C14-HSLs can promote the biofilm formation of *A. ferrooxidans* on the surface of pyrites [26]. Bellenberg et al. reported that the acyl-HSLs addition caused different effects on the pyrite dissolution of *A. ferrivorans*, *Acidiferrobacter sp.*, and *L. ferrooxidans* [24]. Gao et al. found that the *afeI/R* gene cluster overexpression significantly enhanced the adhesion and erosion ability of the cells [47]. A model of the QS system participating in the regulation of the bioleaching progress was proposed [47]. As shown in **Figure 3**, the QS system regulated the EPS synthesis and the biofilm formation, stimulating the planktonic cells to transform to the sessile state. Simultaneously, the sulfur metabolism, bioerosion capacity, and bioleaching efficiency were enhanced by the QS system regulation. The discovery of AfeI/R in regulating the bioleaching process indicated that the QS system plays an important role in regulating the biooxidation process of minerals, and the QS-regulating bioleaching model provides the theoretical basis for studying the control strategy and technologies of acidophilus bacteria in the bioleaching process.

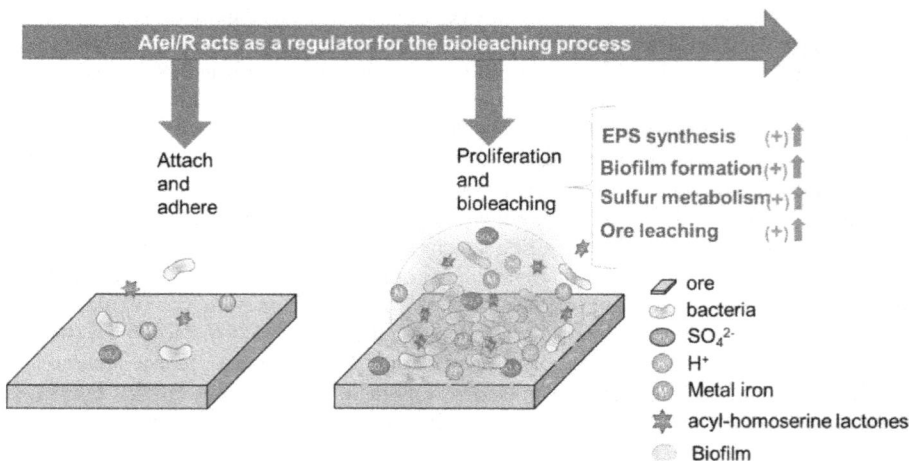

**Figure 3.**
*The regulation of AfeI/R on the bioleaching process of Acidithiobacillus ferrooxidans [47].*

## 6. Conclusion

The regulation function of QS system is an important research content in the study of the co-evolution of microbial community and environment. This chapter systematically describes the QS system in acidophiles including the distribution of QS system, the types of QS system signal molecules, the regulatory function, and application of QS system. Current research shows that the quorum sensing system is involved in the process of cell growth, energy metabolism, the interaction between bacteria and minerals, and the co-evolution process of acidophiles and the extreme environment.

The research of QS system in *A. ferrooxidans* is relatively more extensive than the other acidophiles. Taking *A. ferrooxidans* as an example, the discovery of the energy-dependent regulatory strategy of the AfeI/ R in *A. ferrooxidans* indicated that some chemoautotrophic sulfur-iron-oxidizing bacteria may use the QS system to build the co-evolution process from the response to energy substrates to the regulation on cell growth and population density in the sulfur-and-iron-contained environments. This QS-regulated adaptive strategy may be an important way for chemoautotrophs to adapt to their growth environments and to obtain an ecological competitive advantage.

Due to the complex metabolism and difficulty in the genetic manipulation of acidophiles, the research progress of the QS system in acidophiles has been slow. There is still a lot of room for the research of the QS system in acidophiles. Is there another QS system different from the LuxI/R? In addition to the reported acyl-HSLs type of signal molecules, are there other types of signal molecules in acidophiles? What kind of interspecies regulatory role of the QS system exists in various acidophiles? Moreover, the regulatory functions and molecular mechanisms of the QS system in acidophiles need to be further explored and analyzed. The answers to these questions will not only help to recognize the regulatory functions and mechanisms of the QS system in acidophiles but also help reveal the survival adaptation strategies of microorganisms in extreme environments.

## Acknowledgements

We acknowledge the support of the National Natural Science Foundation of China (32070057).

*Quorum Sensing of Acidophiles: A Communication System in Microorganisms*
DOI: *http://dx.doi.org/10.5772/intechopen.100572*

## Conflict of interest

The authors declare no conflict of interest.

## Author details

Xueyan Gao[1,2*], Jianqiang Lin[1], Linxu Chen[1], Jianqun Lin[1] and Xin Pang[1]

1 State Key Laboratory of Microbial Technology, Shandong University, Qingdao, P.R. China

2 Medical Science and Technology Innovation Center, Shandong First Medical University and Shandong Academy of Medical Sciences, Jinan, P.R. China

*Address all correspondence to: gaoxueyan@sdu.edu.cn

IntechOpen

# References

[1] Baker-Austin C, Dopson M. Life in acid: pH homeostasis in acidophiles. Trends in Microbiology. 2007;**15**: 165-171. DOI: 10.1016/j.tim.2007.02.005

[2] Raquel Quatrini, D.B.J. Acidophiles: Life in Extremely Acidic Environments; 2016

[3] Parsek MR, Greenberg EP. Sociomicrobiology: The connections between quorum sensing and biofilms. Trends in Microbiology. 2005;**13**:27-33. DOI: 10.1016/j.tim.2004.11.007

[4] Miller MB, Bassler BL. Quorum sensing in bacteria. Annual Review of Microbiology. 2001;**55**:165-199. DOI: 10.1146/annurev.micro.55.1.165

[5] Juhas M, Eberl L, Tümmler B. Quorum sensing: The power of cooperation in the world of *Pseudomonas*. Environmental Microbiology. 2005;7:459-471. DOI: 10.1111/j.1462-2920.2005.00769.x

[6] Kai P, Bassler BL. Quorum sensing signal-response systems in Gram-negative bacteria. Nature Reviews. Microbiology. 2016;**14**:576. DOI: 10.1038/nrmicro.2016.89

[7] Fuqua C, Greenberg EP. Listening in on bacteria: Acyl-homoserine lactone signalling. Nature Reviews. Molecular Cell Biology. 2002;**3**:685-695. DOI: 10.1038/nrm907

[8] Asfahl KL, Schuster M. Additive effects of quorum sensing anti-activators on *Pseudomonas aeruginosa* virulence traits and transcriptome. Frontiers in Microbiology. 2017;**8**:2654. DOI: 10.3389/fmicb.2017.02654

[9] Kim BS, Jang SY, Bang YJ, Hwang J, Koo Y, Jang KK, et al. QStatin, a selective Inhibitor of quorum sensing in *Vibrio* species. MBio. 2018;**9**. DOI: 10.1128/mBio.02262-17

[10] Oger P, Kim KS, Sackett RL, Piper KR, Farrand SK. Octopine-type Ti plasmids code for a mannopine-inducible dominant-negative allele of traR, the quorum-sensing activator that regulates Ti plasmid conjugal transfer. Molecular Microbiology. 1998;**27**:277-288

[11] Wang M, Schaefer AL, Dandekar AA, Greenberg EP. Quorum sensing and policing of *Pseudomonas aeruginosa* social cheaters. Proceedings of the National Academy of Sciences of the United States of America. 2015;**112**:2187-2191. DOI: 10.1073/pnas.1500704112

[12] Whiteley M, Diggle SP, Greenberg EP. Progress in and promise of bacterial quorum sensing research. Nature. 2018;**555**:126. DOI: 10.1038/nature25977

[13] Tomasz A. Control of the competent state in *Pneumococcus* by a hormone-like cell product: An example for a new type of regulatory mechanism in bacteria. Nature. 1965;**208**:155-159. DOI: 10.1038/208155a0

[14] Nealson KH, Platt T, Hastings JW. Cellular control of the synthesis and activity of the bacterial luminescent system. Journal of Bacteriology. 1970;**104**:313-322. DOI: 10.1128/jb.104.1.313-322.1970

[15] Eberhard A, Burlingame AL, Eberhard C, Kenyon GL, Nealson KH, Oppenheimer NJ. Structural identification of autoinducer of *Photobacterium fischeri* luciferase. Biochemistry. 1981;**20**:2444-2449. DOI: 10.1021/bi00512a013

[16] Engebrecht J, Nealson K, Silverman M. Bacterial bioluminescence: Isolation and genetic analysis of functions from *Vibrio fischeri*. Cell. 1983;**32**:773-781. DOI: 10.1016/0092-8674(83)90063-6

[17] Engebrecht J, Silverman M. Identification of genes and gene products necessary for bacterial bioluminescence. Proceedings of the National Academy of Sciences of the United States of America. 1984;**81**: 4154-4158. DOI: 10.1073/pnas.81.13.4154

[18] Wang R, Lin J-Q, Liu X-M, Pang X, Zhang C-J, Yang C-L, et al. Sulfur oxidation in the acidophilic autotrophic *Acidithiobacillus spp*. Frontiers in Microbiology. 2019;**9**. DOI: 10.3389/fmicb.2018.03290

[19] Rohwerder T, Gehrke T, Kinzler K, Sand W. Bioleaching review part A: Progress in bioleaching: Fundamentals and mechanisms of bacterial metal sulfide oxidation. Applied Microbiology and Biotechnology. 2003;**63**:239. DOI: 10.1007/s00253-003-1448-7

[20] Olson GJ, Brierley JA, Brierley CL. Bioleaching review part B: progress in bioleaching: Applications of microbial processes by the minerals industries. Applied Microbiology and Biotechnology. 2003;**63**:249-257. DOI: 10.1007/s00253-003-1404-6

[21] Farah C, Vera M, Morin D, Haras D, Jerez CA, Guiliani N. Evidence for a functional quorum-sensing type AI-1 system in the extremophilic bacterium *Acidithiobacillus ferrooxidans*. Applied and Environmental Microbiology. 2005;**71**:7033-7040. DOI: 10.1128/aem.71.11.7033-7040.2005

[22] Rivas M, Seeger M, Holmes DS, Jedlicki E. A Lux-like quorum sensing system in the extreme acidophile *Acidithiobacillus ferrooxidans*. Biological Research. 2005;**38**:283-297

[23] Gao XY, Fu CA, Hao L, Gu XF, Wang R, Lin JQ, et al. The substrate-dependent regulatory effects of the AfeI/R system in *Acidithiobacillus ferrooxidans* reveals the novel regulation strategy of quorum sensing in acidophiles. Environmental Microbiology. 2020. DOI: 10.1111/1462-2920.15163, 10.1111/1462-2920.15163

[24] Bellenberg S, Diaz M, Noel N, Sand W, Poetsch A, Guiliani N, et al. Biofilm formation, communication and interactions of leaching bacteria during colonization of pyrite and sulfur surfaces. Research in Microbiology. 2014;**165**:773-781. DOI: 10.1016/j.resmic.2014.08.006

[25] Rivas M, Seeger M, Jedlicki E, Holmes DS. Second acyl homoserine lactone production system in the extreme acidophile *Acidithiobacillus ferrooxidans*. Applied and Environmental Microbiology. 2007;**73**:3225. DOI: 10.1128/aem.02948-06

[26] Gonzalez A, Bellenberg S, Mamani S, Ruiz L, Echeverria A, Soulere L, et al. AHL signaling molecules with a large acyl chain enhance biofilm formation on sulfur and metal sulfides by the bioleaching bacterium *Acidithiobacillus ferrooxidans*. Applied Microbiology and Biotechnology. 2013;**97**:3729-3737. DOI: 10.1007/s00253-012-4229-3

[27] Scholtissek A, Ullrich SR, Mühling M, Schlömann M, Paul CE, Tischler D. A thermophilic-like ene-reductase originating from an acidophilic iron oxidizer. Applied Microbiology and Biotechnology. 2017;**101**:609-619. DOI: 10.1007/s00253-016-7782-3

[28] Christel S, Herold M. Multi-omics reveals the lifestyle of the acidophilic, mineral-oxidizing model species *Leptospirillum ferriphilum* (T). Applied and Environmental Microbiology. 2018;**84**. DOI: 10.1128/aem.02091-17

[29] Bellenberg S, Buetti-Dinh A. Automated microscopic analysis of metal sulfide colonization by acidophilic microorganisms. Applied and Environmental Microbiology. 2018;**84**. DOI: 10.1128/aem.01835-18

[30] Parsek MR, Greenberg EP. Acyl-homoserine lactone quorum sensing in Gram-negative bacteria: A signaling mechanism involved in associations with higher organisms. Proceedings of the National Academy of Sciences of the United States of America. 2000;**97**: 8789-8793. DOI: 10.1073/pnas.97.16.8789

[31] Teplitski M, Eberhard A, Gronquist MR, Gao M, Robinson JB, Bauer WD. Chemical identification of N-acyl homoserine lactone quorum-sensing signals produced by *Sinorhizobium meliloti* strains in defined medium. Archives of Microbiology. 2003;**180**:494-497. DOI: 10.1007/s00203-003-0612-x

[32] Ahlgren NA, Harwood CS, Schaefer AL, Giraud E, Greenberg EP. Aryl-homoserine lactone quorum sensing in stem-nodulating photosynthetic bradyrhizobia. Proceedings of the National Academy of Sciences of the United States of America. 2011;**108**:7183-7188. DOI: 10.1073/pnas.1103821108

[33] Soulère L, Guiliani N, Queneau Y, Jerez CA, Doutheau A. Molecular insights into quorum sensing in *Acidithiobacillus ferrooxidans* bacteria via molecular modelling of the transcriptional regulator AfeR and of the binding mode of long-chain acyl homoserine lactones. Journal of Molecular Modeling. 2008;**14**:599-606. DOI: 10.1007/s00894-008-0315-y

[34] Chabert N, Bonnefoy V, Achouak W. Quorum sensing improves current output with *Acidithiobacillus ferrooxidans*. Microbial Biotechnology. 2017;**11**:136. DOI: 10.1111/1751-7915.12797

[35] Díaz M, San Martin D. Quorum sensing signaling molecules positively regulate c-di-GMP effector PelD encoding gene and PEL exopolysaccharide biosynthesis in extremophile bacterium

*Acidithiobacillus thiooxidans*. Genes. 2021;**12**. DOI: 10.3390/genes12010069

[36] Mamani S, Moinier D, Denis Y, Soulere L, Queneau Y, Talla E, et al. Insights into the quorum sensing regulon of the acidophilic *Acidithiobacillus ferrooxidans* revealed by transcriptomic in the presence of an acyl homoserine lactone superagonist analog. Frontiers in Microbiology. 2016;**7**:1365. DOI: 10.3389/fmicb.2016.01365

[37] Lee J, Toshinari M, Hong SH, Woodl Thomas K. Reconfiguring the quorum-sensing regulator SdiA of Escherichia coli to control biofilm formation via indole and N-acylhomoserine lactones. Applied and Environmental Microbiology. 2009

[38] Donlan RM. Biofilms: Microbial life on surfaces. Emerging Infectious Diseases. 2002;**8**:881-890. DOI: 10.3201/eid0809.020063

[39] Tan CH, Koh KS, Xie C, Tay M, Zhou Y, Williams R, et al. The role of quorum sensing signalling in EPS production and the assembly of a sludge community into aerobic granules. The ISME Journal. 2014;**8**:1186-1197

[40] Kozlova EV, Khajanchi BK, Sha J, Chopra AK. Quorum sensing and c-di-GMP-dependent alterations in gene transcripts and virulence-associated phenotypes in a clinical isolate of *Aeromonas hydrophila*. Microbial Pathogenesis. 2011;**50**:213-223

[41] Sharma IM, Petchiappan A, Chatterji D. Quorum sensing and biofilm formation in mycobacteria: Role of c-di-GMP and methods to study this second messenger. IUBMB Life. 2014;**66**:823-834. DOI: 10.1002/iub.1339

[42] Ueda A, Wood TK. Connecting quorum sensing, c-di-GMP, pel polysaccharide, and biofilm formation in *Pseudomonas aeruginosa* through

tyrosine phosphatase TpbA (PA3885).
PLoS Pathogens. 2009;**5**:e1000483. DOI:
10.1371/journal.ppat.1000483

[43] Solano C, Echeverz M, Lasa I.
Biofilm dispersion and quorum sensing.
Current Opinion in Microbiology.
2014;**18**:96-104. DOI: 10.1016/j.
mib.2014.02.008

[44] Banderas A, Guiliani N.
Bioinformatic prediction of gene
functions regulated by quorum sensing
in the bioleaching bacterium
*Acidithiobacillus ferrooxidans*.
International Journal of Molecular
Sciences. 2013;**14**:16901-16916. DOI:
10.3390/ijms140816901

[45] Gehrke T, Telegdi J, Thierry D,
Sand W. Importance of extracellular
polymeric substances from *Thiobacillus
ferrooxidans* for bioleaching. Applied
and Environmental Microbiology.
1998;**64**:2743-2747. DOI: 10.0000/
PMID9647862

[46] Harneit K, Göksel A, Kock D,
Klock JH, Gehrke T, Sand W. Adhesion
to metal sulfide surfaces by cells of
*Acidithiobacillus ferrooxidans*,
*Acidithiobacillus thiooxidans* and
*Leptospirillum ferrooxidans*.
Hydrometallurgy. 2006;**83**:245-254.
DOI: 10.1016/j.hydromet.2006.03.044

[47] Gao X-Y, Liu X-J, Fu C-A, Gu X-F,
Lin J-Q, Liu X-M, et al. Novel strategy
for improvement of the bioleaching
efficiency of *Acidithiobacillus
ferrooxidans* based on the AfeI/R
quorum sensing system. Minerals.
2020;**10**:222

# *Acidithiobacillus* Its Application in Biomining Using a Quorum Sensing Modulation Approach

*Juan Carlos Caicedo and Sonia Villamizar*

## Abstract

A group of particular acidophiles microorganisms (bacteria and archaea) known as chemolithoautotrophs are capable of using minerals as fuel. Its oxidation generates electrons to obtain energy and carbon that is obtained by fixing $CO_2$ from the air. During this aerobic mineral oxidation, metals are solubilized or biodegraded. Metal bioleaching usually is used in biomining and urban biomining approaches to recovery metals such as copper, gold and zinc. Several species of bacterial genus *Acidithiobacillus* display a great bioleaching activity. Bacterial attachment and biofilm formation are the initial requirements to begin a successful bioleaching process. Biofilm formation in *Acidithiobacillus* bacteria is strongly regulated by cell to cell communication system called Quorum Sensing. The goal of this chapter is to review the Quorum Sensing system mediated by the autoinducer N-acyl- homoserine-lactones in the Bacterium *Acidiothiobacillus ferroxidans*, in order to enhance and to boost the bioleaching technologies based in the use of this bacterium. The main applications of the cell-to-cell communication system concepts in *A. ferrooxidans* are reviewed in this chapter. It is that the addition of synthetic autoinducers molecules, which act as agonist of quorum sensing system, especially those with long acyl chains, both as single molecules ($C_{12}$-AHL, 3-hydroxy-$C_{12}$-AHL, $C_{14}$-AHL, and 3-hydroxy-$C_{14}$-AHL) or as a mixture ($C_{14}$-AHL/3- hydroxy-$C_{14}$-AHL/3-oxo-C14-AHL) increased the adhesion to sulfur and pyrite and enhance the metal bioleaching in urban biomining approaches.

**Keywords:** AfeI/R, Biofilm, Bioleaching, EPS, Autoinducer, Synthetic Agonist

## 1. Introduction

As the world population grows, the reduction of several natural sources becomes more evident. One of the major concerns is the fastest decrease of metals ores. Currently, the hyper-technological society demands several metals in order to attend the rising request for electrical and electronical devices. The fact of the production of these devices is gradually cheapest and its useful life is increasingly shorter. Equally, the advertising campaigns increase the people desires to renew the older devices by a brand-new much faster and modern. It has been triggering the generation of waste of electrical and electronical devices (E-waste). It has been rising 5 folds in the last 20 years [1]. The organic and inorganic fractions of E-waste could be a serious threat to environment and public health if its final disposition is not done accurately. The incineration and landfill are the global strategies frequently applied in a whole world for final disposal of E-waste [2]. Although of its highly toxic nature, E-waste particularly, printed circuit boards (PCBs) are a promising secondary source of metals. The concentrations of copper (Cu) and precious metals, such as gold (Au), platinum (Pt) and palladium (Pd), are high compared to natural mines [3]. Urban E-waste mining knowing as Biomining is expected to be an important secondary source of metals in the near future. PCBs are electronic components whose most abundant metal is Cu (roughly 20–25% by weight). PCBs are also composed of a substantial amount of precious metals. Precious metals constitute the largest fraction of the value of discarded PCBs and are the main economic driver of metal recovery [4]. In the last 20 years, great efforts have been focused in the recovery of these metals from the E-waste, mainly based in traditional approaches such as pyrometallurgy and hydrometallurgy. However, these two approaches have been inconvenient, due to the high consumption of energy and the high emission of polluting gases in the first one approach and in the second one the high generation of acid leachates, these leachates could reach easily the subterranean and surface water bodies [5].

Recent biotechnological developments have made possible the adaptation of various microorganisms to hostile bioleaching environments. Thus, emerging a new approach for the metal recovery from E-waste called biohydrometallurgy. This process involve microorganism in order to make use of metal elements for their structural and metabolic functions. This process based on the use of biomass has the advantages over pyrometallurgy and hydrometallurgy processes such as: it does not require the construction of a huge infrastructure, it is not necessary to deal with a large amount of harmful residual pollutants generated in the process. However, the main bottleneck that prevents this process from being attractive at an industrial level despite its high efficiency, it is the drop in yield of metal recovery when using large amounts of E-waste. However, the main bottleneck that prevents this process from being attractive at an industrial level despite its high efficiency, it is the drop in yield of metal recovery when using large amounts of E-waste. The cell-to-cell communication system known as Quorum sensing, it allows bacteria to colonize new ecological niches, to resist environmental changes and toxic substances, enhances its competitiveness and to resist host defenses [6]. As this quorum sensing system regulates near to 25% of non-essential genes, its knowledge, understanding and modulation could help to enhance the bioleaching reactions at industrial level. In this chapter, we focused to review the quorum sensing system in the bacterium *Acidithiobacillus ferrooxidans*, which bacterium is widely recognized by its tremendous bioleaching ability, and discussed the application potential of QS of this bacterium in bioleaching.

## 2. *Acidiothibacillus ferroxidans*

The gram-negative bacterium *A. ferrooxidans* belongs to a gamma-proteobacterium group. *A. ferrooxidans* is a facultative anaerobe bacterium, which is able to grow from oxic to anoxic environments. *A. ferrooxidans* is acidophile, mesophilic and chemolithoatotrophic, that obtain energy mainly by the oxidation of ferrous iron ($Fe^{2+}$), elemental and inorganic sulfur and other sulfur compounds [7]. *A. ferrooxidans* grows optimally at temperature of around 30°C and pH below 2. This bacterium is a natural inhabitant of ecological niches associated to pyritic ores, coal deposits and acid drainages [8]. Another exceptional aspect of A. *ferrooxidans* is its outstanding ability to thrive in environments with higher concentration of dissolved $Fe^{2+}$ around of $10^{16}$ folds superior that neutral environments. Opposing to what would expect that this massive soluble Fe concentration leads to bacterial DNA and protein damage, *A. ferrooxidans* uses Fe as micronutrient and energy source, which makes it an exceptional model microorganism for the homeostasis and assimilation of Fe [9].

The genome size of *A. ferrooxidans* is 2.98 MB with 58.7% of G + C, a total of 3217 protein encoding genes of which 2070 have a putative function [10]. Like others bioleaching bacteria, *A. ferrooxidans* must face with huge heavy metal concentration in their natural niches. Diverse gene cluster have been related with the tolerance to mercury, arsenic and copper. These genes including: copper extrusion system, resistance-nodulation cell division family transporters and encoding copper translocating ATPases [10]. Two Quorum Sensing systems mediated by the autoinducer acyl homoserine lactones molecules (AHLs) have been described in *A. ferrooxidans,* the first one AfeI/R resembling to classical *LuxI/R* system presents in numerous gram-negative bacteria [11] and the second one called *act* for de acyl transfer function also produces AHL autoinducer molecules [12]. The review of these quorum sensing systems in *A. ferrooxidans*, its conformation likewise, the biological traits regulated by these communication systems such as: energy metabolism, attachment, biofilm production and toxic tolerance are the main purpose of this chapter.

## 3. Bioleaching reaction by *A. ferrooxidanss*

In the bacterium *Acidithiobacillus ferrooxidans*, bioleaching is carried out by three types of mechanisms: (i) contact, (ii) independent contact and (iii) cooperative. These three mechanisms are not mutually exclusive and generally work synergistically. The contact mechanism is characterized by the imperative need for the bacterial production of Extracellular Polymeric Substances (EPS). This EPS is essential for the subsequent formation of biofilms that will facilitate the contact between the bacterial wall and the surface of the E-waste and thus achieve the oxidation of the metal.

Previous studies have shown that *A. ferroxidans* bacteria do not adhere randomly to solid surfaces, but there is still a fine chemotaxis mechanism involved in preferential adherence to metals [13]. In the independent contact mechanism, the bacteria are not attached to the surface and the oxidation of $Fe^{2+}$ in solution to $Fe^{3+}$ occurs. Fe will subsequently act as an oxidizing agent, solubilizing the metals contained in the E-waste particles [14]. The cooperative mechanism is a combination of the two previous mechanisms in which the attached bacteria and the oxidizing agent $Fe^{3+}$ in solution cooperate to oxidize the metals present in the E-waste particles. In the E-waste, specifically in PCI, metals are not present in the form of metallic sulphides. These are present as zero valence metals such as $Cu^0$, $Zn^0$, $Ni^0$, etc. Thus, the ferric ions and/or protons produced biologically from ferrous ions or the reduced sulfur compounds are responsible for the conversion of insoluble neutral metals to water soluble metallic ions.

## 4. Quorum sensing systems in extremophiles and its biotechnological applications

Cooperative bacterial behavior is one of the most intensively studied topics of microbial ecology in recent decades. The understanding of this lifestyle allows us to revel the fitness strategies of bacteria to thrive in very different environmental niches. Quorum sensing (QS) is a system of bacterial cell–cell communication that enables the microorganism to sense a minimum number of cells (quorum) in order to respond to external stimuli in a concerted fashion. QS system is based on the production, releasing to the extracellular environment and perception of small diffusible molecules known as auto-inducers (AI) [15]. Thus, this communication allows the bacteria to detect changes in the density of the population, which leading to generate variations in the concentration of nutrients, oxygen, inorganic molecules, pH and osmolarity in the extracellular environment. In this way, an increase in the bacterial population causes the accumulation of the AI molecule in the medium, which, when detected by the bacteria, generates changes in the expression of several target genes in order to regulate phenotypes traits in pathogenic or environmental bacteria. The main traits regulated are: attachment, virulence, resistance and bioluminescence among others [16].

In Gram-negative bacteria including the extremophiles (i.e. thermophiles, halophiles, psychrophiles and acidophiles) the most studied system involves AI molecules of the N-acyl homoserine lactones (AHLs) or autoinducer-1 (AI-1) type. This QS system is involved in intra-species and inter-species communications [17]. It was first described in the bioluminescent marine bacterium *Vibrio fischeri*. This system is considered as the QS paradigm in Gram-negative bacteria, which includes at least four elements: (i) AHLs as signal molecule or AI-1, (ii) an AHL synthase protein responsible for the synthesis of AI-1, (iii) a transcriptional regulator (belonging to the R protein family) and (iv) a cis-active palindromic DNA sequence that is the target of the R-AHL binary complex [18]. The general mechanism of action of AHLs molecules is to diffuse into the target cell and bind to an R-type transcriptional regulator to generate its dimerization (**Figure 1**). This occurs when the concentration of AHLs in the extracellular medium is adequate to generate a gradient towards the intracellular space [19]. The concentration of AHL molecules in the extracellular environment not only depends on their synthesis, but also on their diffusion, transport and degradation. The diffusion of AHL molecules depends on their size.

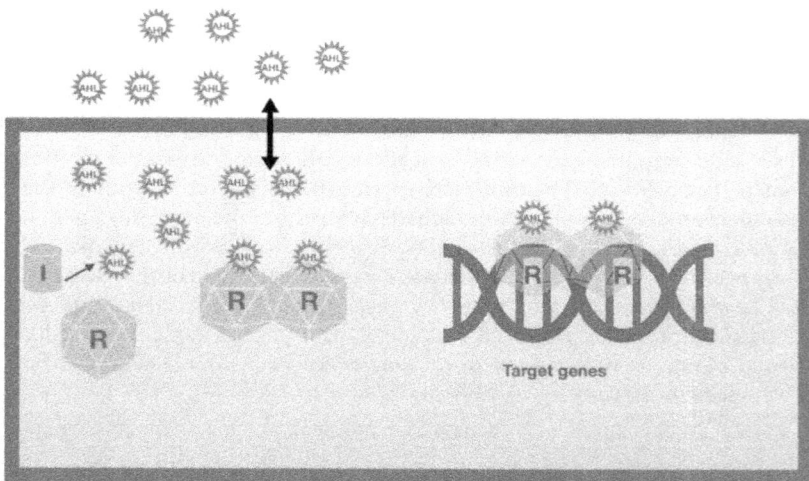

**Figure 1.**
*Schematic representation of Quorum Sensing system type I AI-1. AHLs are synthesized inside the bacteria by protein I. The AHLs molecules diffuse towards the extracellular environment, increasing their concentration both in the internal and external environment of the cell in proportion to the increase in the cell population. Upon reaching the threshold concentration, the AHLs molecules bind with the R protein, which dimerizes and promotes or represses the transcription of target genes. Some of these target genes have a binding site for the R/AHL complex in their promoter region.*

Short acyl chain AHLs diffuse freely across the cell membrane [20], while long chain AHLs are excreted by transporters [21].

The biological meaning of cell to cell communication in extreme environments is not completely comprehended, however, this communication systems play a pivotal role in the survival to fitness to these restrictive habitats. However, a better understanding about the adaptation strategies regulated by quorum sensing in these extreme ecological niches, could possibly contribute to the design and development of innovative approaches for the biotechnological solutions with industrial, medical and research applications.

Most of the Quorum sensing systems present in mesophyll bacteria are shared by thermophilic bacteria. Bioinformatic analysis revealed the existence of complete AI-2 systems in 17 thermophilic bacteria from phyla Deinococcus-Thermus.

For other hand, 18 show incomplete QS systems having only LuxS [22]. Particularly interesting is the AI-2 production in the bacteria *Thermotoga maritima* and *Pyrococcus furiosus*. These two hyperthermophiles bacteria produce AI-2 signals despite lacking of synthase enzyme LuxS [23]. *T. maritima* and *P. furiosus* have genes that encode for the SAH (S-adenosylhomocysteine) hydrolase, which catalyzes the cleavage of SAH producing adenosine and homocysteine. Then, the adenosine produced is transformed to AI-2. This production of AI-2 in independent fashion of luxS could be explained by the action of the rearrangement of phosphorylated ribose, which rearrangement seems to be modulated by temperature depending mechanism [24].

Psychrophile bacteria produce DiKetoPiperazine (DKPs) molecules, which can activate the LuxR circuit. *Pseudoalteromonas haloplanktis* bacterium produce eight different types of DKPs [25]. DKPs shows a great potential for disruption of the AHLs quorum sensing system in the pathogenic bacteria *Burkholderia cenocepacia* and *Pseudomonas aeuroginosa*. It results from the direct inhibition of AHL synthase. This fact represents a great potential for the alternative treatment of disease such as cystic fibrosis [26]. Another industrial application mediated by QS in psychrophile bacteria is the alginate overproduction at low temperature by the *Pseudomonas mandelii*. Pseudomonads bacteria produce alginate by the expression of genes *algU* and *mucA*. These genes expression are QS regulated [27]. In the biofilm formation process at low temperature (4–15°C) the alginate operon transcription is keeping active because its repressor is downregulated. Biofilm formation shows to be an adaptation strategy for survival at low temperatures [28].

Acidophiles bacteria have endowed with Quorum Sensing systems such as: AI-1, AI-2 and CAI-1. The genome of acidophilic iron oxidizing bacterium *Ferrovum* sp. harbors a thermophilic ene-reductase (ERs). The ERs Family enzymes has shown that it could play a role in the Quorum Sensing mediation of oxidative stress response [29]. *Ferrovum* sp. lacks of encoding gene for the AHL synthase LuxI. The gene *foye-1* in the genome of *Ferrovum* is located directly downstream of the LuxR. A previous study suggests that ERs FOYE-1 in this bacterium could be responsible for the production of HLA and triggers the LuxR perception [30]. The ene-reductase enzymes have increasing relevance in the field of bio-catalysis because of its superior stability. This fact rises the biotechnological potential of these ERs.

Enteric pathogenic bacterium *Vibrium cholerae* produces and senses three AIs such as: AI-1, DPO and CAI-1. The first two are recognized for the interspecies communication role, while the last one is employed as intra-genus communication AI. These all three QS systems work together for the transition from an aerobic to an anaerobic environment. This QS represses the virulence and biofilm formation in the *Vibrium cholerae* bacterium at high cell density [31].

The bacterium *Acidithiobacilus ferrooxidans* widely recognized by its bioleaching ability displays two quorum sensing systems, (i) a completely functional

LuxI/R like, encoding the gene cluster *afeI-orf3-afeR* [32]. *At. ferrooxidans* is able to produce nine types of AHLs with acyl chains, whose length ranges between 8 and 16 carbons. A recent study shows that depending on energy substrate, the protein AfeI synthetizes different types of acyl-HSLs, similarly, the regulation of different metabolic systems also depends on substrate energy [33]. (ii) The second Quorum Sensing system in *A. ferrooxidans* has been identified using a reporter strain strategy. This QS system was termed *act* QS system. This reporter strain evidenced the production of HLA molecule e.g. $C_{14}$ acyl-HSL by *E. coli* cloned with the *act* gene [34]. The quorum sensing system *act* is mainly expressed when *A. ferrooxidans* cells are grown in culture medium enriched with $Fe^{+2}$ than when the cells are cultivated in microbiological medium enriched with sulfur. The role of *act* Quorum Sensing system of *A. ferrooxidans* remains uncertain.

Synthetic biology is based on the use of engineering principles to design and apply new biological components, besides to integrate functions and traits into the present ones in order to standardize or modulate their behavior. The completely understanding of a particular Quorum sensing circuit enable to design tools to precisely and predictably manipulate cell responses under quorum sensing regulation. That is, QS and synthetic biology research have been highly complementary, with QS research expanding the synthetic biology toolkit and synthetic biology providing new tools for investigating QS [35]. The HLA QS system is a relatively simple system to synthetic biology approaches, due to few components to comprise the circuit. Several studies using synthetic biology to engineer cell were performed to modulate behavior when the phenotype is depending on cell density. One of the most usual approaches in this case is the manipulation of the regulator protein LuxR i.e. modifying its affinity to is cognate or not cognate AHL signals, regulating the transcription to control the number of copies of LuxR and by modifying the AHL binding site [36, 37]. Some of the above mentioned approaches were used with success in the biomining sewage bioremediation [38].

## 5. Improving of bioleaching ability through EPS regulation synthesis in *A. ferrooxidans*

Secretion of EPS (Extracellular Polymeric Substances) by *A. ferrooxidans* is a determinant factor for biofilm formation and an essential feature for mineral dissolution in bioleaching process [39]. Bioleaching denotes direct and indirect actions exerted by microorganisms that lead to the dissolution of metals in ores or urban biomining strategies. It has been widely demonstrated that bacterial attachment to various mineral surfaces and the formation of a well-established biofilm, contribute significantly to enhancing bioleaching activity, since, the biofilm allows the formation of reaction space identified as "surface conditioning layer" between the ores or E-waste and bacteria [40]. A previous urban bioleaching study was done using a partition system based on a semipermeable layer, in order to avoid the contact between bacteria and E-waste. This study showed that the lack of bacterial attachment reduced 25% the recovery of copper [41]. These physiological steps are mediated by extracellular polymeric substances (EPS), which are composed of polysaccharides, proteins, and lipids. Furthermore, by increasing EPS secretion and biofilms formation on mineral surfaces, attached cells extend their reactive space and obtain significantly higher amounts of substrate than planktonic cells [42]. Thus, the increase of EPS secretion could increase bioerosion capacity of the overexpression strain on the substrates and enhance the efficiency of the bioleaching process.

Early studies showed that it is feasible to modulate *A. ferrooxidans* adhesion to pyrite particles by addition of synthetic AHL and HLSs analogues. Nevertheless, the questionable point of this study was that the results were obtained using an indirect approach, i.e. the number of the residual planktonic cells was employed in order to calculate the number of cells attached to the surface of pyrite [43]. Subsequent studies based on the use of different microscopy techniques provided conclusive evidence on the effect of bacterial adhesion on bioleaching efficiency [44]. In the bioleaching process, a bacterial attachment to copper sulfide minerals occurs selectively [45]. Thus, in the cell population that interacts electrostatically or hydrophobically with pyrite or sulfur surfaces, an increase in cell densities is observed on these surfaces. It triggers the activation of Quorum sensing regulon in those cells that remaine attached to solid surfaces much faster than in planktonic cells. This activation could contribute to favor the biofilm producer phenotype at the cell population. As it was mentioned above, the AfeI/R Quorum Sensing system promotes the EPS production and consequently contributes to the biofilm formation. Based on these observations, several studies have been implemented the addition of synthetic auto-inducers in order to make the bioleaching reaction more efficient and get a higher yield of metals. The QS agonists employed were principally those with Carbon long length e.g. $C_{12}$-$C_{14}$ [46]. In a previous study using direct microscopy technics, the researchers reported that AHLs with long acyl chains used as single molecules ($C_{12}$-AHL, 3-hydroxy-$C_{12}$-AHL, $C_{14}$-AHL, and 3-hydroxy-$C_{14}$-AHL) or as a mixture ($C_{14}$-AHL/3- hydroxy-$C_{14}$-AHL/3-oxo-C14-AHL) increased the adhesion to sulfur and pyrite [47]. Though, studies focused on elucidating and clearly understating all molecular steps produced by the addition of different AHLs autoinducers, and its effects on biofilm formation are required.

## 6. Conclusions

The cell communication system Quorum sensing is a ubiquitous phenomenon in prokaryotes. This communication system modulates near to 25% of no essential genes. Induction or quenching of quorum sensing system could be a high valuable tool to select the desire trait in bacteria. The bacterium *A. ferrooxidans* is equipped with a fully functional Quorum Sensing system AfeI/R, which promotes the EPS secretion and biofilm formation. These traits are extremely desirable for the bioleaching reaction both for natural ores and urban biomining processes. However, special care must be taken in order to choose the energy source for the media culture where the bacteria will be grown and the bioleaching reaction will take place. As mentioned before, long chain synthetic agonists of AHL Quorum sensing system could favor the biofilm formation when the Sulfur is provided as energy source, otherwise, the addition of AHL synthetic agonist of long chain in microbiological media culture using iron as energy source, reduce drastically the bacterial growth and repress the genes responsible for the EPS and biofilm formation. Act, the second Quorum system in *A. ferrooxidans* has shown an over expression cultured in the media enriched with iron. Nevertheless, the mechanistic details about its encoding genes and their biological roles remain unclear. In addition to modulating the biofilm formation, AfeI/R Quorum Sensing system in *A. ferrooxidans* controls the grown rate, the metabolic systems and membrane permeability. Finally, the synergism between synthetic biology and Quorum sensing research enables a wide specter of approaches, in order to modulate the behavior of acidophile microorganisms with potential applications in the biomining industry.

## Acknowledgements

The author thanks to all researchers of department of microbiology and biotechnology at Gujarat University for their unavailable help and disposition. This work was supported by Administrative Department of Science and Technology MINCIENCIAS and Energy Mining Planning Unit UPME, Grant No 80740-008-2020.

## Conflict of interest

The authors declare no conflict of interest.

## Author details

Juan Carlos Caicedo[1*] and Sonia Villamizar[2]

1 Universidad de Santander, Faculty of Exact, Natural and Agricultural Science, Bucaramanga, Colombia

2 School of Agricultural and Veterinarian Sciences, São Paulo State University (UNESP), Jaboticabal, Brazil

*Address all correspondence to: jua.caicedo@mail.udes.edu.co

IntechOpen

# References

[1] Robinson, B.H., 2009. E-waste: an assessment of global production and environmental impacts. Sci. Total Environ. 408, 183-191 Doi: doi. org/10.1016/j.scitotenv.2009.09.044.

[2] Baldé, C.P., Wang, F., Kuehr, R., Huisman, J., 2015. The Global E-Waste Monitor 2014, United Nations University, IAS – SCYCLE, Bonn, Germany

[3] Hadi, P., Xu, M., Lin, C.S.K., Hui, C.W., McKay, G., 2015. Waste printed circuit board recycling techniques and product utilization. J. Hazard. Mater. 283, 234-243. doi:10.1016/j. jhazmat. 2014.09.032.

[4] Cui, J., Zhang, L., 2008. Metallurgical recovery of metals from electronic waste: A review. J. Hazard. Mater. 158, 228-256. doi:10.1016/j. jhazmat.2008.02.001.

[5] Birloaga,I.;DeMichelis,I.;Ferella,F.;B uzatu,M.;Veglio,F. 2013,. Study on the influence of various factors in the hydrometallurgical processing of waste printed circuit boards for copper and gold recovery. Waste Manag. 33, 935-941

[6] Ng WL, Bassler BL, 2009. Bacterial quorum-sensing network architectures. Annual Review of Genetics 43, 197-222.

[7] Hedrich, S., and Johnson, D. B. (2013). Aerobic and anaerobic oxidation of hydrogen by acidophilic bacteria. FEMS Microbiol. Lett. 349, 40-45. doi: 10.1111/1574-6968.12290

[8] Gonzalez-Toril E, Llobet-Brossa E, Casamayor EO, Amann R, Amils R: Microbial ecology of an extreme acidic environment, the Tinto River. Appl Environ Microbiol 2003, 69(8):4853-4865.

[9] Quatrini R, Jedlicki E, Holmes DS: Genomic insights into the iron uptake mechanisms of the biomining microorganism Acidithiobacillus ferrooxidans. J Ind Microbiol Biotechnol 2005, 32(11-12):606-614.

[10] Valdés, J., Pedroso, I., Quatrini, R. et al. Acidithiobacillus ferrooxidans metabolism: from genome sequence to industrial applications. BMC Genomics 9, 597 (2008). https://doi.org/10. 1186/1471-2164-9-597

[11] Schaefer, A.L., Greenberg, E.P., Oliver, C.M., Oda, Y., Huang, J.J., Bittan-Banin, G., et al. (2008) A new class of homoserine lactone quorum-sensing signals. Nature 454: 595-599. https://doi.org/10.1038/nature07088.

[12] Rivas, M., Seeger, M., Jedlicki, E., and Holmes, D.S. (2007) Second acyl homoserine lactone production system in the extreme acidophile Acidithiobacillus ferrooxidans. Appl Environ Microbiol 73: 3225. https://doi. org/10.1128/aem. 02948-06.

[13] Rawlings DE (2002) Heavy metal mining using microbes. Annu Rev Microbiol 56:65-91

[14] Sand W, Gehrke T, Hallmann R, Schippers A (1995) Sulfur chemistry, biofilm, and the (in)direct attack mechanism—a critical evalua- tion of bacterial leaching. Appl Microbiol Biotechnol 43:961-966

[15] Caicedo, J. C. et al. (2016.) Bacteria from the citrus phylloplane can disrupt cell-cell signalling in Xanthomonas citri and reduce citrus canker disease severity. Plant Pathology, v. 65, n. 5, p. 782-791, https://doi.org/10.1111/ ppa.12466

[16] Kai, P.; Bassler, B.L. (2016 )Quorum sensing signal-response systems in Gram-negative bacteria. Nat. Rev. Microbiol., 14, 576.

[17] Whitehead N.A., Barnard A.M., Slater H., Simpson N.J. and Salmond G.P. 2001. Quorum sensing in Gram-negative bacteria. FEMS Microbiol. Rev. 25: 365-404.

[18] Fuqua W. C., Winans S. C. and Greenberg E. P. 1994. Quorum sensing in bacteria: the LuxR-LuxI family of cell density-responsive transcriptional activators. J. Bacteriol. 176:269-75.

[19] Nilsson P., Olofsson A., Fagerlind M., Fagerström T., Rice S., Kjelleberg S. and Steinberg P. 2001. Kinetics of the AHL regulatory system in a model biofilm system: how many bacteria constitute a "quorum"? J. Mol. Biol. 309(3): 631-640.

[20] Kaplan H.B and Greenberg E.P. 1985. Diffusion of autoinducer is involved in regulation of the Vibrio fischeri luminescense system. J. Bacteriol. 163: 1210-1214.

[21] Chan Y.Y., Bian H.S., Chin Tan T.M., Mattmann M.E., Geske G.D., Igarashi J., Hatano T., Suga H., Blackwell H.E. and Chua K.L. 2007. Control of quorum sensing by a Burkholderia pseudomallei multidrug efflux pump. J. Bacteriol. 189: 4320-4324.

[22] Rivas, M., Seeger, M., Holmes, D.S., and Jedlicki, E. (2005) A Lux-like quorum sensing system in the extreme acido- phile Acidithiobacillus ferrooxidans. Biol Res 38: 283. https://doi.org/10.4067/S0716-97602005000200018.

[23] Kaur, A., Capalash, N., Sharma, P., 2018. Quorum sensing in thermophiles: prevalence of autoinducer-2 system. BMC Microbiol. 18, 62.

[24] Nichols, J.D., Johnson, M.R., Chou, C.J., Kelly, R.M., 2009. Temperature, not LuxS, mediates AI-2 formation in hydrothermal habitats. FEMS Microbiol. Ecol. 68, 1173-1181.

[25] Amandeep Kaur, Neena Capalash, Prince Sharma, (2019).Communication mechanisms in extremophiles: Exploring their existence and industrial applications, Microbiological Research,Volume 221, doi.org/10.1016/j.micres.2019.01.003.

[26] Mitova, M., Tutino, M.L., Infusini, G., Marino, G., De Rosa, S., (2005). Exocellular peptides from antarctic psychrophile Pseudoalteromonas haloplanktis. Mar. Biotechnol. 7, 523-531.

[27] Buroni S., Scoffone V. C., Fumagalli M., Makarov V., Cagnone M., Trespidi G., et al. (2018). Investigating the mechanism of action of diketopiperazines inhibitors of the Burkholderia cenocepacia quorum sensing synthase cepi: a site-directed mutagenesis study. Front. Pharmacol. 9:836. 10.3389/fphar.2018.00836

[28] Fazli, M., Almblad, H., Rybtke, M.L., Givskov, M., Eberl, L., Tolker-Nielsen, T., 2014. Regulation of biofilm formation in Pseudomonas and Burkholderia species. Environ. Microbiol. 16, 1961-1981.

[29] Vasquez-Ponce, F., Higuera-Llanten, S., Pavlov, M.S., Ramirez-Orellana, R., Marshall, S.H., Olivares-Pacheco, J., 2017. Alginate overproduction and biofilm formation by psychrotolerant Pseudomonas mandelii depend on temperature in Antarctic marine sediments. Electron. J. Biotechnol. 28, 27-34.

[30] Toogood, H.S., Gardiner, J.M., Scrutton, N.S., 2010. Biocatalytic reductions and chemical versatility of the old yellow enzyme family of flavoprotein oxidoreductases. Chem. Cat. Chem. 2, 892-914.

[31] Scholtissek A, Ullrich SR, Mühling M, Schlömann M, Paul CE, Tischler D. A thermophilic-like ene-reductase originating from an acidophilic iron oxidizer. Appl

Microbiol Biotechnol. 2017 Jan;101(2):609-619. doi: 10.1007/s00253-016-7782-3.

[32] Bridges AA, Bassler BL. 2019. The intragenus and interspecies quorum-sensing autoinducers exert distinct control over Vibrio cholerae biofilm formation and dispersal. PLoS Biol 17:e3000429. doi.org/10.1371/journal.pbio.3000429

[33] Farah C., Vera M., Morin D., Haras D., Jerez C.A. and Guiliani N. 2005. Evidence for a functional quorum-sensing type AI-1 system in the extremophilic bacterium Acidithiobacillus ferrooxidans. Appl. Environ. Microbiol. 71:7033-40

[34] Gao, X.-Y., Fu, C.-A., Hao, L., Gu, X.-F., Wang, R., Lin, J.-Q., Liu, X.-M., Pang, X., Zhang, C.-J., Lin, J.-Q. and Chen, L.-X. (2021), The substrate-dependent regulatory effects of the AfeI/R system in Acidithiobacillus ferrooxidans reveals the novel regulation strategy of quorum sensing in acidophiles. Environ Microbiol, 23: 757-773. https://doi.org/10.1111/1462-2920.15163

[35] Stephens K, Bentley WE. Synthetic Biology for Manipulating Quorum Sensing in Microbial Consortia. Trends Microbiol. 2020 Aug;28(8):633-643. doi: 10.1016/j.tim.2020.03.009. Epub 2020 Apr 24. PMID: 32340782.

[36] Wang, B. et al. (2015) Ampli!cation of small molecule-inducible gene expression via tuning of intracellular receptor densities. Nucleic Acids Res. 43, 1955-1964

[37] Shong, J. and Collins, C.H. (2013) Engineering the esaR promoter for tunable quorum sensing- dependent gene expression. ACS Synth. Biol. 2, 568-575

[38] Hong, S. H., Hegde, M., Kim, J., Wang, X., Jayaraman, A., and Wood, T.

K. (2012). Synthetic quorum- sensing circuit to control consortial biofilm formation and dispersal in a microfluidic device. Nat. Commun 3, 613.

[39] Rivas, M., Seeger, M., Jedlicki, E., and Holmes, D.S. (2007) Second acyl homoserine lactone production system in the extreme acidophile *Acidithiobacillus ferrooxidans*. Appl Environ Microbiol 73: 3225 doi.org/10.1128/aem.02948-06.

[40] Gao, X.-Y.; Liu, X.-J.; Fu, C.-A.; Gu, X.-F.; Lin, J.-Q.; Liu, X.-M.; Pang, X.; Lin, J.-Q.; Chen, L.-X. Novel Strategy for Improvement of the Bioleaching Efficiency of Acidithiobacillus ferrooxidans Based on the AfeI/R Quorum Sensing System. Minerals 2020, 10, 222. https://doi.org/10.3390/min10030222

[41] Liu, H.L.; Chen, B.Y.; Lan, Y.W.; Cheng, Y.C. (2003). SEM and AFM images of pyrite surfaces after bioleaching by the indigenous *Thiobacillus thiooxidans*. Appl. Microbiol. Biotechnol. 62, 414-420

[42] Silva RA, Park J, Lee E, Park J, Choi SQ, Kim H (2015) Influence of bacterial adhesion on copper extraction from printed circuit boards. Sep Purif Technol 143:169-176

[43] Yang, H.; Luo, W.; Gao, Y. Effect of Acidithiobacillus ferrooxidans on humic-acid passivation layer on pyrite surface. Minerals 2018, 8, 422.

[44] Ruiz LM, Valenzuela S, Castro M, Gonzalez A, Frezza M, Soulere L, Rohwerder T, Sand W, Queneau Y, Jerez CA, Doutheau A, Guiliani N (2008) AHL communication is a widespread phenom- enon in biomining bacteria and seems to be involved in mineral- adhesion efficiency. Hydrometallurgy 94:133-137

[45] González A, Bellenberg S, Mamani S, Ruiz L, Echeverría A,

Soulère L, Doutheau A, Demergasso C, Sand W, Queneau Y, Vera M, Guiliani N. AHL signaling molecules with a large acyl chain enhance biofilm formation on sulfur and metal sulfides by the bioleaching bacterium Acidithiobacillus ferrooxidans. Appl Microbiol Biotechnol. 2013 Apr;97(8):3729-37. doi: 10.1007/s00253-012-4229-3.

[46] Echeverría A, Demergasso C (2009) Assessment of microbial adhesion in mixed cultures to sulfide minerals using CARD-FISH techni- ques. Adv Mat Res 71-73:83-86

[47] Stevens AM, Queneau Y, Soulère L, Von Bodman S, Doutheau A (2011) Mechanisms and synthetic modulators of AHL- dependent gene regulation. Chem Rev 111:4-27

Chapter 6

# Immunomodulatory Potential of *Lactobacillus acidophilus*: Implications in Bone Health

*Asha Bhardwaj, Leena Sapra, Bhupendra Verma and Rupesh K. Srivastava*

## Abstract

*Lactobacillus acidophilus* is homofermentative anaerobic rod-shaped gram-positive bacteria. *L. acidophilous* is one of the most common probiotics and is used for the treatment of various gastrointestinal, metabolic and inflammatory disorders. *L. acidophilous* produces antimicrobial compounds, maintains gut permeability and prevents dysbiosis. *L. acidophilus* also shows various other properties such as: it is anticarcinogenic, lowers serum cholesterol level and improves lactase metabolism of host. One of the most significant property of *L. acidophilous* is that it modulates the immune system and can prevent various inflammatory disorders. *L. acidophilous* influences several immune cells such as Th17 cells and Tregs. Various studies reported that inflammation induces bone loss and leads to several bone pathologies such as osteoporosis, rheumatoid arthritis and periodontitis. Recent studies have shown the potential of probiotics in preventing inflammation mediated bone loss. *L. acidophilous* is one of these probiotics and is found capable in inhibition of various bone disorders. *L. acidophilous* restores the dysregulated immune homeostasis and prevents inflammatory bone loss. Thus, *L. acidophilous* can be a potential therapeutic for the management of various bone pathologies. In this book chapter we reviewed various immunomodulatory properties of *L. acidophilous* along with its efficacy in preventing dysbiosis and maintaining gut permeability. We also discussed the potential role of *L. acidophilous* as a therapeutic for the management of inflammation induced bone disorders.

**Keywords:** probiotics, *Lactobacillus acidophilus*, immune cells, dysbiosis, gut permeability, bone

## 1. Introduction

The word "Probiotics" is derived from Latin language meaning life [1] and came into attention in 1953 by the German scientist Werner Kollath who defined them as "active substances that are essential for healthy development of life". Later on, in 1992 Fuller defined them as "a live microbial feed supplement which beneficially affects the host animal by improving its intestinal microbial balance" [2]. Currently probiotics are defined as "live organisms that when administered in adequate amounts confer health benefits on the host" and are specified by the Food and Agriculture Organization of the United Nations and the World Health Organization

(FAO/WHO, 2001). Probiotics are present mainly in fermented foods like cheese, bread, wine, kefir and kumis and are commercially available in the market as powders, tablets and packets [1]. Probiotics are used from centuries for the treatment of various diseases but it was not known until the 20th century that probiotics are healthy bacteria that replace harmful microbes in the gut and regulate gut flora [3]. The most extensively used probiotics are *Lactobacillus* and *Bifidobacterium*. Other common probiotics are *Bacillus*, *Streptococcus*, *Enterococcus* and the fungus *Saccharomyces* [4]. Probiotics are used for the treatment of various gastrointestinal disorders like irritable bowel syndrome (IBS), inflammatory bowel disease (IBD), infectious diarrhea, *Clostridium difficile* colitis and antibiotic associated diarrhea and many other metabolic disorders such as obesity, diabetes and non-alcoholic fatty liver disease [5, 6]. Several mechanisms are involved in preventive activities of probiotics such as they modulate the immune system, regulate gut barrier and protect from pathogens [7]. For a probiotic to be successful it should have various qualities like it should be resistant to the low pH present in gastrointestinal tract, able to colonize in the gut, adhere to the epithelium and be able to activate the immune system. It should also have several other qualities such as it should be of human origin, non-pathogenic, noncariogenic and influence the local metabolic activity [4]. *Lactobacillus acidophilus* (LA) is one of the most common probiotics and is present in several commercially available food products and dietary supplements [8]. LA exhibits antimicrobial, anticarcinogenic and anti-inflammatory properties [8, 9]. LA has various properties that make it a good probiotic such as it is acid tolerant, bile tolerant, has lactase activity, can adhere to the human epithelial cells, lowers serum cholesterol level, prevents infection, modulates immune response, improves lactose metabolism of host, etc. [10]. Several commercially accessible strains of LA have probiotics ability like LA-1 to LA-5 (Chr. Hansen, Demark), NCFM (Dansico, Madison), SBT-2026 (Snow brand milk products, Japan), DDS-1 (Nebraska cultures, Nebraska), etc. LA NCFM is the most common LA strain and is regarded safe by the US Food and Drug Administration (FDA) [10]. LA has immunomodulatory properties and is considered for the treatment of various inflammatory diseases such as IBD, cancer, etc. [11, 12]. Use of probiotics for the prevention of bone loss has recently gain much attention. Probiotics prevent osteoporosis and other bone diseases like arthritis and periodontitis by influencing the immune system or via other mechanisms. Various studies have shown the potential of LA in preventing bone diseases [9]. Bone disorders like osteoporosis, rheumatoid arthritis (RA) and periodontitis are immune disorders and it is observed that LA has potential of preventing these disorders by modulating the immune system. Thus, LA can act as a therapeutic for the treatment of various bone fragilities.

In this chapter we summarized some of the mechanisms which are responsible for the health promoting effects of LA focussing primarily on immunomodulatory properties of LA. We also discussed the role of LA in preventing inflammatory bone loss and how modulation of gut microbiota and maintenance of gut integrity by LA can play a role in regulating bone health.

## 2. *Lactobacillus acidophilus*

LA is a type of lactic acid bacteria (LAB). LAB constitute a group of gram-positive, acid tolerant, catalase negative, non-sporulating and generally rod-shaped bacteria [13] that are frequently associated with dairy, meat and plants [14]. LAB produce lactic acid from carbohydrate fermentation which make them important in fermentation and agriculture-based industries. They are used for imparting unique textures and flavors and for preservation and acidification of different food items [10]. LAB comprised a number of genera such as *Lactobacillus*, *Enterococcus*, *Streptococcus*, *Cornobacterium*, *Leuconostoc*, *Lactococcus*, *Bifidobacterium* and *Sporolactobacillus* which are further subdivided into species, subspecies, variants and strains [15, 16]. *Lactobacillus* is the largest genus of lactic acid bacteria having more than 145 species [17]. *Lactobacilli* is part of human microbial flora which colonizes in the human gastrointestinal and urinary tract [18]. *Lactobacillus* species are the first ones to colonize the gut after birth where they provide various health promoting effects. *Lactobacillus* species have various qualities that make them suitable as probiotics. They are resistant to stomach pH and bile juices, can adhere to the mucosa, inhibit growth of other harmful bacterial species and have immunomodulatory properties [19]. *Lactobacilli* encompass a wide range of species that have role in various biochemical and physiological functions [10]. LA is one of the most known species belonging to *Lactobacillus* genus. LA was earlier named as *Bacillus acidophilus* and first isolated in 1900 from the human infant feces by Moro [19]. Almost 80% of the yogurts in America have LA [19]. LA is rod shaped homofermentative anaerobic having size of approximatively 2–10 μM. LA is a thermophile and grows optimally at a temperature of 37 to $45^0$ C and at pH range of 4–5 [16]. Highest growth is observed at pH between 5.5 and 6.0 whereas growth ceases at pH 4. Diet is one of the major source of LA in gut. Various commercially available food products such as yogurt and milk are supplemented with LA due to its probiotic value [19]. LA is part of human microbiota and is isolated from digestive, oral and vaginal areas but Claesson's characterization revealed that gastrointestinal tract is its main environment [19]. It is observed that LA supplementation to humans in heat killed form is completely safe. It is observed that heat killed LA provides protection to immunodeficient mice infected with *Candida albicans* [20]. Simakachorn et al. also reported that LA supplementation induces no adverse effect in children having diarrhea [21]. LA is found effective in the treatment of various inflammatory disorders like IBD, diabetes, cancer, etc. [19]. LA prevents these inflammatory disorders by regulating the immune homeostasis. Therefore, the immunomodulatory potential of LA can be used for the management of various disorders. From recent studies it is observed that LA also inhibits inflammatory bone loss and can prevent various bone fragilities such as osteoporosis, RA and periodontitis. Below we reviewed the various immune modifying properties of LA. Later in the chapter, we discussed the role of immune modification by LA in prevention of inflammation induced bone loss. Apart from immunomodulation LA also prevents dysbiosis and increase in gut permeability which are discussed later in the chapter.

### 2.1 LA role in modulating the immune system

LA has great immunomodulatory properties. Because of the immune modifying properties of LA it is considered for the treatment of various inflammatory diseases. LA can be an inexpensive therapeutic for treatment of numerous clinical manifestations involving malfunctioning of the immune system. Here we discuss various studies highlighting the importance of LA as a potential therapeutic for the prevention of several immune related disorders. It is observed that LA and *L. plantarum* supplementation for 60 days enhanced the expression of genes related to innate immune response in crayfish [22]. Feeding of probiotic "dahi" consisting of LA and *Bifidobacterium bifidum* reversed decrease in immune response in aging mice [23]. LA and *Bifidobacterium animalis* subspecies lactis decreased inflammation of intestinal epithelial cells by modifying the toll like receptor 2 (TLR2) mediated Nuclear Factor kappa-light-chain-enhancer of activated B cells (NF-κB) and mitogen-activated protein kinase (MAPK) signaling pathways [24]. Feeding of milk fermented with LA and *L. casei* increased both the phagocytic and lymphocytic activity in swiss mice [25]. LA strain NCFM increased gram-positive immune response in *C. elegans* by modulating key immune signaling pathways such as p38 MAPK and β-catenin signaling pathways [26]. LA can be used for the prevention of obesity related effects. LA-KCTC3925 supplementation significantly attenuated the levels of splenic and hepatic cyclooxygenase-2 (COX-2) mRNA expression and intracellular adhesion molecule-1 (ICAM-1) expression in high fat diet induced obese mice [27]. LA supplementation generated non-specific immune response in germ free mice [28].

LA regulates the secretion of cytokines from various immune cells and maintains the balance between inflammatory and anti-inflammatory cytokines. It is observed that LA treatment significantly altered the production of interleukin (IL)-4 and interferon (IFN)-γ from splenocytes in the presence of purified tumor antigen. LA and *L. reuteri* modulated the cytokine response in neonatal gnotobiotic pigs infected with human rotavirus (HRV). LA and *L. reuteri* treatment in HRV infected pigs significantly enhanced the production of Th1 and Th2 cytokine responses as indicated by the higher concentration of IL-12, IL-10, IL-4 and INF-γ in these pigs. Treated pigs also had higher concentration of transforming growth factor (TGF)-β as compared to the controls. Thus, LA and *L. reuteri* supplementation can maintain immune homeostasis by regulating TGF-β level after HRV infection [29]. LA induced the production of cytokines such as IL-1β, TNF-α, IL-10 and IFN-γ from human peripheral blood mononuclear cells (PBMCs) [30]. LA significantly downregulated the production of anti-inflammatory cytokines and reactive oxygen species (ROS) whereas increased the production of anti-inflammatory cytokines from PBMCs isolated from Parkinson's disease patients [31]. Chen et al. showed that LA suppressed IL-17 production in experimental colitis model by suppressing expression of IL-23 and TGF-β1 and downstream phosphorylation of phospho-signal transducer and activator of transcription 3 (p-STAT3) [11]. It is observed that LA downregulated the expression of IL-1α, IL-1β, IL-8, monocyte chemoattract protein (MCP)-1 and C-X-C motif ligand 3 (CXCL3) in bovine mammary epithelial (BME) cells after Lipopolysaccharide (LPS) challenge. LA also increased the expression level of negative regulators of TLRs viz. toll interacting protein, ubiquitin-editing enzyme A20 and single immunoglobulin IL-1 single receptor in BME cells after LPS challenge [32]. Thus, LA can be a treatment option for bovine mastitis which is characterized by inflammation of the mammary glands. LA treatment significantly enhanced the expression of IL-1β, IFN-α, IFN-γ, interferon regulatory factor (IRF)-7, interferon-inducible transmembrane protein M3 and 2′,5′-oligoadenylate synthetase in chicken macrophages in response to avian influenza virus [33]. LA induced the production of TGF-β and inflammatory cytokines such as IL-6 and tumour necrosis factor (TNF-α) in dendritic cells (DCs) cocultured with intestinal epithelial cells [34]. It is observed that administration of LA strain L36 to germ-free mice induced higher expression of cytokines associated with Th2 cells such as IL-6, IL-5 and TGF-β and Th17 cells like IL-17A, IL-6 and TNF-α [35]. In dextran sodium sulphate (DSS) induced colitis LA administration suppressed the production of pro-inflammatory cytokines such as IL-6, TNF-α and IL-17 in colon tissues. *In vitro* LA treatment stimulated Tregs and the production of IL-10 [36]. LA treatment downregulated the expression of inflammatory cytokines, chemokines and myeloperoxidase in mice model of DSS induced colitis [37]. LA treatment also restored the number of colon goblet cells by inducing IL-10 expression and suppressing proinflammatory cytokines expression in DSS induced colitis [38]. LA treatment induced various antiviral cytokines and chemokines such as IL-1β, regulated upon activation, normal T cell expressed and presumably secreted (RANTES), macrophage colony stimulating factor (MCSF), eotaxin and IFN-α in lung and IL-17 in peyer's patches of influenza virus infected mice [39].

One of the mechanisms by which LA inhibits the progression of inflammatory diseases is by modulating T cells (**Figure 1**). It is observed that LA-CGMCC 7282 along with *C. butyricum* CGMCC 7281 exerts strong anti-inflammatory effects and can prevent Th1 and Th2-type ulcerative colitis [40]. LA protected the β-lactoglobulin sensitized mice by reducing the allergic inflammation [41]. LA treatment was found to be positively associated with decreased mRNA expression of IL-17 and RORγt and reduced proliferation of Th17 cells under both *in vitro* and *in vivo* models of β-lactoglobulin allergy [41]. In case of HRV infection it is observed that varied doses of LA induced different effects. Wen et al. showed that low dose of LA enhanced IFN-γ producing T cell response but decreased Tregs response the production of TGF-β and IL-10 from Tregs. On the other hand, higher dose of LA upregulated Tregs response in gnotobiotic pigs infected with HRV [42].

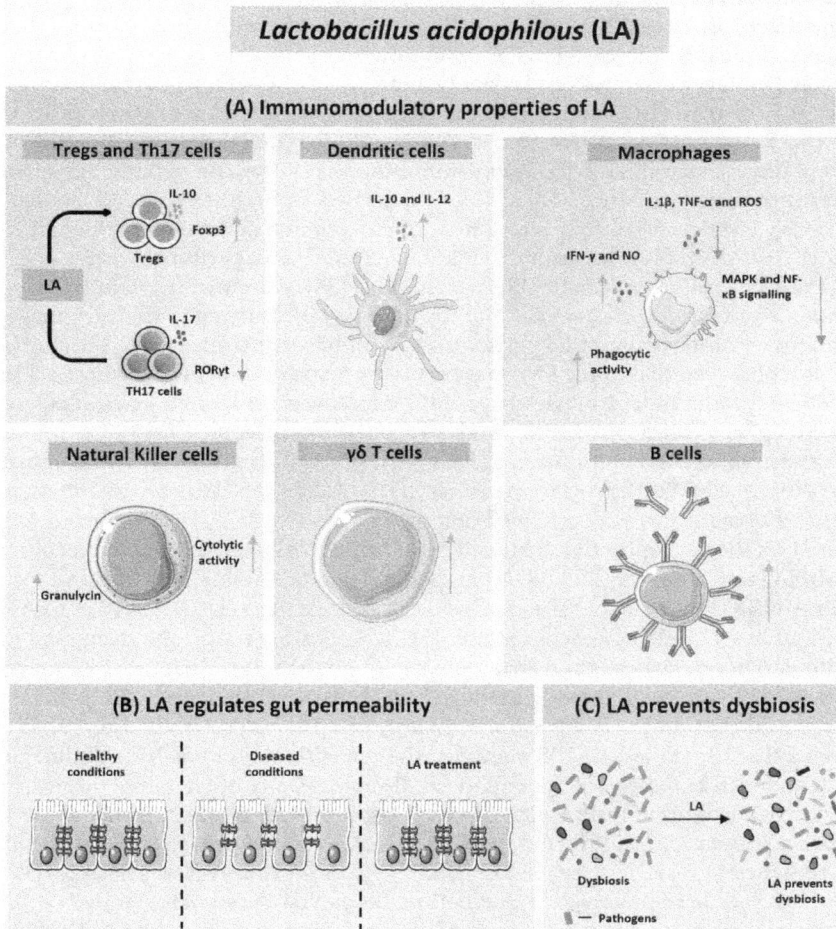

**Figure 1.**
*Schematic diagram depicting immumodulatory properties and effect of LA on gut permeability and dysbiosis. (A) LA influence the activity of various immune cells such as Tregs and Th17 cells, dendritic cells, macrophages, natural killer cells, γδ T cells and B cells. (B) LA prevents the increase in gut permeability which leads to various diseases such as IBD, IBS, etc. (C) LA restores gut microbiota composition in dysbiotic conditions.*

LA strain L-92 attenuated the progression of 2, 4-dinitroflurobenzene induced contact dermatitis by regulating Tregs in spleen and cervical lymph nodes. LA-L-92 administration also enhanced FoxP3, IL-10 and TGF-β levels as compared to the controls [43]. LA and *B. longum* administration to the colitis mice model upregulated the number of Tregs and γδ T cells in intraepithelial lymphocytes [44]. Li et al. showed that LA prevents β-immunoglobulin allergy by regulating the balance the Tregs/Th17 cells and activation of TLR2/NF-Kb signaling pathways [45]. It is observed that LA lysates administration in DSS induced colorectal cancer mice model suppressed macrophage (type 2 i.e. M2) polarization, increased the number of CD8$^+$ T cells and effector memory T cells and decreased the number of Tregs in tumor microenvironment [12]. It is reported that when LA is administered after saline challenge in pigs, it increased the number of leucocytes and CD4$^+$ T cells whereas when challenged with LPS decreased the number of both CD4$^+$ and CD8$^+$ T lymphocytes, leukocytes, expression of IL-6 and TNFα as compared to the control diet. LA modulates the activity of other immune cells also. LA enhanced the production of IL-10 and IFN-γ from splenocytes induced with concanavalin A (Con A) and significantly increased the phagocytic activity of peritoneal macrophages [46]. Surface layer protein (Slp) isolated from LA-NCFM reduced the production of IL-1β, TNF-α and ROS in LPS induced RAW 264.7 cells via suppression of MAPK and NF-κB signaling pathways. Slp also attenuated the production of nitic oxide (NO) and prostaglandin E2 (PGE$_2$) by inhibiting the expression of inducible nitric oxide synthase (iNOS) and COX-2 [40]. SLP derived from the LA-CICC6074 also like LA-NCFM decreased the secretion of TNF-α and enhanced the secretion of NO in RAW 264.7 cells [47]. LA stimulated M2 macrophages in peritoneal cavity and Tregs and Th2 cells in spleen of DSS treated mice [38]. Administration of LA, *L. rhamnosus* and *Bifidobacterium lactis* to the mice significantly enhanced the phagocytic activity of leukocytes and peritoneal macrophages as compared to the controls [48]. It is observed that non-LPS component of LA strain DSS-1 induced the IL-1α and TNF-α production by macrophages [49]. Moreover, LA treated macrophages showed higher expression of IFN-γ and costimulatory molecule CD40 [33]. LA induces activation and maturation of DCs [50]. LA stimulated the IFN-β response in DCs in a myeloid differentiation primary response 88 (Myd88) dependent manner [51]. Konstantinov et al., showed that major SlpA of LA-NCFM interacted with the DCs via their receptor dendritic cell-specific intercellular adhesion molecule-3-grabbing non-integrin (DC-SIGN) and modulated the function of DCs and T cells [52]. LA-NCFM upregulated the expression of defense genes in DCs such as IL-12 and IL-10 in a TLR-2 dependent manner [51]. LA administration also decreased the degranulation of mast cells and eosinophils [53]. It is observed that pre-treatment with LA-L-92 enhanced the natural killer (NK) cells activity in lung [39]. L-92 also reduced the number of neutrophils in lung of influenza treated mice [39]. It is reported that heat killed LA 205 increased the cytolytic activity of NK cells in a time and dose dependent manner. LA enhanced the cytotoxicity of NK cells by elevating the expression of granulysin which is cytolytic granule component in NK cells [54]. In elderly population administration of LA and *Bifidobacterium bifidum* enhanced the frequency of B cells in peripheral blood [55]. LA prevents various diseases by modulating the level of antibody production. LA can be beneficial in preventing food allergy. It is observed that LA-AD031 and *Bifidobacterium lactis* ADO11 administration significantly decreased the ovalbumin (OVA) specific IgE, IgA and IgG1 in OVA and cholera toxin sensitized mice [53]. Oral administration of heat killed LA attenuated hypersensitivity responses in bovine β-lactoglobulin sensitized mice model. LA administration decreased inflammatory cells and IgE production. Along with IgE production LA treatment enhanced mRNA expression levels of

CD25, FoxP3 and TGF-β whereas decreased the expression of IL-17A, RORγt and IL-10 in allergic group [56]. Intermediate dose of LA increased rotavirus specific IgA and IgG antibody secreting cells and memory B cells in response to rotavirus vaccine. Thus, LA administration can be used to improve the efficacy of rotavirus vaccine and thus can be effective against rotavirus diarrhea [57]. LA increase the IgA, IL-10 and IFNγ producing cells in small intestine [58]. LA improved the immunogenicity of Newcastle Disease vaccines (NDV). Chicks treated with both LA and vaccine have increased IgG and NDV antibody titres than the only vaccinated group [59]. Feeding of probiotic "dahi" (curd) containing LA and *L. casei* ameliorated the secretory IgA and lymphocyte proliferation in *Salmonella enteritidis* infected mice [60]. These probiotics also increased the proliferative response of splenocytes to LPS and con A [48]. Su et al. showed that LA SW1 could function as a promising immune adjuvant in DNA vaccine against foot and mouth disease (FMD). Oral LA-SW1 enhanced the levels of anti-FMDV antibody titres and FMDV neutralizing antibodies [61]. LA lysates also increased the antitumor activity of CTLA-4 monoclonal antibody [58]. LA not only promotes immune response but also inhibits unnecessary lethal immune responses. It is observed that LA along with *B. bifidus* or *B. infantis* suppressed the mitogen activated cell proliferation of splenocytes and PBMCs and arrested the cell cycle at G0/G1 phase. At higher concentrations these probiotics inhibited the mitogen activated overactive immune response and at lower concentration skewed the balance of Th1/Th2 balance towards Th1 [62].

On the basis of these above discussed studies, we can consider that LA has immune regulatory properties. LA has capabilities of regulating various innate and adaptive immune cells and therefore can maintain immune homeostasis. Altogether, these studies suggest that immunomodulatory properties of LA can be employed for regulating the disrupted immune homeostasis in various inflammatory diseases.

## 2.2 LA role in preventing dysbiosis

Trillions of microbes reside in human gastrointestinal tract. These microbes contribute in number of vital functions related to health. These are source of essential vitamins and nutrients. Microbes extract energy from food, modulate immune system and maintain gut permeability. Gut microbiota usually promotes human health but alteration in the gut lead to various clinical manifestations. Alteration in gut microbiota is termed as dysbiosis. Gut microbiota can be altered by various factors like diet, toxins, pathogens, drugs, antibiotics, etc. [63]. Dysbiosis is reported in various diseases like IBS, IBD, diabetes, obesity, cardiovascular diseases, asthma, allergy, etc. [63–67]. Dysbiosis is also observed in leukemia where selective modulation of *Lactobacillus* species is reported [68]. Dysbiosis is the reason for various vaginal diseases like aerobic vaginitis, bacterial vaginosis and vulvovaginal candidiasis. In vagina of reproductive aged women microbial homeostasis is maintained by the mutualistic relationship between microbes and the host which provide protection against vaginal infections by preventing the colonization of opportunistic pathogens [69]. LA, *L. iners* and *L. crispatus* are the most abundant bacterial species in vaginal tract [70]. Role of LA in preventing dysbiosis is reported by various studies. LA-DSS-1 administration improved the abundance of beneficial bacteria like *Lactobacillus* spp. and *Akkermansia* spp. in caecum [71]. In ulcerative colitis patients, supplementation of LA, *Lactobacillus salivarius* and *Bifidobacterium bifidus* along with anti-inflammatory drug mesalazine prevented intestinal dysbiosis [72]. LA also decreased dysbiosis and inflammation induced by *Salmonella typhimurium* infection in Th1 and Th2 biased mice [73]. LA reversed the alterations in the gut microbiota composition caused due to administration of high fat diet in animals [74]. Oral administration of LA along with cobiotic ginger extract encapsulated in calcium-alginate beads modulated gut microbiota and prevented 1,2 dimethylhy-drazine (DMH)/DSS induced colitis and precancerous lesions in rats [75]. Probiotic combination consisting of LA, *L. helveticus*, *L. gassari*, *L. crispatus* and *L. salivarius* prevented vaginal dysbiosis by restoring the altered vaginal microbiota to normal level. Probiotics combination enhanced the abundance of *Lactobacillus* while decreased the abundance of *Enterobacter* and *Enterococcus* [76]. Antibiotics use provide protection against wide number of pathogens but also disturb the intestinal microflora balance. On the contrary LA is found to be capable of restoring intestinal microbial homeostasis. It is observed that LA prevent the dysbiosis induced by antibiotic Azithromycin [77]. Synbiotic consisting of inulin, LA, *L. plantarum* W21, *L. lactis* and *Bifidobacterium lactis* W51 prevented stress induced dysbiosis and thus can be useful in preventing dysbiosis induced in stress related diseases like IBS and IBD [78]. LA administration is found effective in treatment of dyspepsia caused by dysbiosis [79]. It is observed that the oral intake of LA-GLA-14 and *L. rham-nosus* HN001 mixture along with bovine lactoferrin prevent vaginal dysbiosis and improve vaginal health. After oral ingestion both the LA-GLA-14 and *L. rhamnosus* HN001 colonize and restore the vaginal microbiota [69]. LA supplementation in mice increases short chain fatty acid (SCFA) producing bacteria and thus decreases the gram-negative bacteria [80].

### 2.3 LA role in maintaining gut permeability

Gut barrier is very important for the regulation of the immune homeostasis and for preventing the access of pathogens into the gut lumen. Through the leaky gut, pathogens invade into the lumen and lead to uncontrolled inflammation. Gut barrier is regulated by tight Junctions (TJs) which are present between the intestinal epithelial cells. TJs are transmembrane proteins and are divided into four groups: claudins, occludin, tricullin and junctional adhesion molecules (JAMs). Transmembrane TJs are linked with the actin cytoskeleton through the cytosolic scaffold proteins like zona occludens (ZOs) which are of three types ZO-1, ZO-2 and ZO-3 [81]. Alteration in expression of TJs leads to increase in gut permeability and intestinal inflammation which is responsible for various inflammatory diseases like IBD [82–86], colon cancer [87] and RA [88].

Several studies have shown that LA administration maintain gut permeability. It is observed that the mixture of LA-KLDS1.0901 and *L. plantarum* KLDS1.0344 prevents chronic alcohol liver injury in mice by improving the gut permeability. *Lactobacillus* mixture inhibits the increase in gut permeability and reduces the abundance of gram-negative bacteria resulting in decrease of LPS entering the portal vein thereby suppressing alcohol promoted liver inflammation [80]. LA along with *L. rhamnosus* and *B. bifidumi* prevented high fat diet induced increase in gut permeability and LPS translocation [74]. LA in combination with ginger extract restored colonic permeability in DMH-DSS induced colon cancer in Wistar rats [75]. Conditioned media of LA significantly prevented the increase in IL-1β induced increase in gut permeability. Conditioned media of LA inhibits IL-1β stimulated decrease in occludin and increase in claudin-1 expression and thus preserve intestinal permeability by normalizing the expression of occludin and claudin-1 [89]. Probiotic combination of LA, *L. reuteri*, *L. casei*, *Streptococcus thermophiles* and *Bifidobacterium bifidum* significantly reduced diabetes incidence and gut permeability [90]. Administration of probiotics LA and *Bifidobacterium infantis* to the pregnant women daily from embryonic day 15- to 2-week-old postnatally maintained the intestinal integrity of preweaned offspring. Thus, LA supplementation to pregnant women can promote barrier function of developing offsprings [91]. LA and *Streptococcus thermophilus* enhanced the barrier function of epithelial cells and protected the epithelial cells from infection induced by enteroinvasive *E. coli* by limiting its adhesion and invasion [92].

## 3. Bone remodeling

Bone is a dynamic and metabolically active organ that is being remodeled throughout the life of the organisms. Bone remodeling is regulated by three different types of bone cells viz. osteoclasts (bone eating cells), osteoblasts (bone forming cells) and osteocytes. Osteoblasts are derived from multipotent stem cells that also give rise to fibroblasts, adipocytes, chondrocytes and myoblasts [93]. Osteoblasts are responsible for formation of osteoid matrix by depositing collagen which later on get calcified. The major constituent of osteoid matrix is type 1 collagen which provide resistance against fractures. Osteoid matrix also consists of various other non-collagenous proteins which are responsible for various critical functions of bone [94]. Osteoblast differentiation depends on a number of paracrine and transcription factors such as Runx2 and osterix and members of bone morphogenetic protein (BMP) family [94]. Osteoclasts are giant polynuclear cells that have special ability of resorbing bone [95]. Osteoclasts differentiate from monocytic progenitors that also give rise to cells of other monocytic lineages such as macrophages, dendritic cells, granulocytes and microglia. Osteoclast differentiation depends on two important cytokines: MCSF and receptor activator of nuclear factor κb ligand (RANKL). MCSF stimulate proliferation and differentiation of osteoclast progenitors. RANKL acts with the help of its receptor RANK and coupling molecule TNF receptor associated factor 6 (TRAF 6) to promote the differentiation and commitment of precursor cells [95]. Bone resorption starts when osteoclasts attach to the surface of bone and form a unique structure called sealing zone. Sealing zone permit the osteoclasts to form resorption space. Osteoclasts acidifies the resorption space to degrade the mineral and organic compartment of the bone. For this osteoclast secrete various lysosomal enzymes such as cathepsin K into the resorption space. To mediate bone resorption osteoclasts, form a specialized structure called as ruffled border that increase the surface area for active transport of $H^+$ through proton pump. Osteoclasts comparatively resorb a large area of bone and then die by apoptosis [94]. Osteocytes are osteoblasts that have been entrapped in the osteoid matrix during matrix calcification under the influence of bone specific alkaline phosphatase produced by osteoblasts [93, 94]. Osteocytes sense mechanical force and tissue strain and send signal to the other osteocytes and osteoblasts by forming cellular network termed as canaliculi permeating the entire bone matrix [93, 94]. Dynamic equilibrium between the osteoblasts and osteoclasts maintains bone integrity. Multiple interactions take place between the bone forming osteoblasts and bone resorbing osteoclasts to regulate the process of bone remodeling [96]. Osteoblasts positively regulate osteoclast differentiation by secreting RANKL and MCSF at pre-osteoblastic stage and negatively by secreting the RANKL decoy receptor osteoprotegerin (OPG). Bone remodeling restore microdamages and ensure the release of calcium and phosphorus in normal host physiology [97]. Bone remodeling consist of four phases viz. activation phase, resorption phase, reversal phase and formation phase [97]. In activation phase, MCSF and RANKL induce the differentiation of osteoclast progenitors into osteoclasts. During resorption phase pre-osteoclasts migrate at the surface of bone and get differentiated into mature osteoclasts and start resorbing bone. Resorption phase is followed by reversal phase where mononuclear cells remove the collagen remnants and prepare

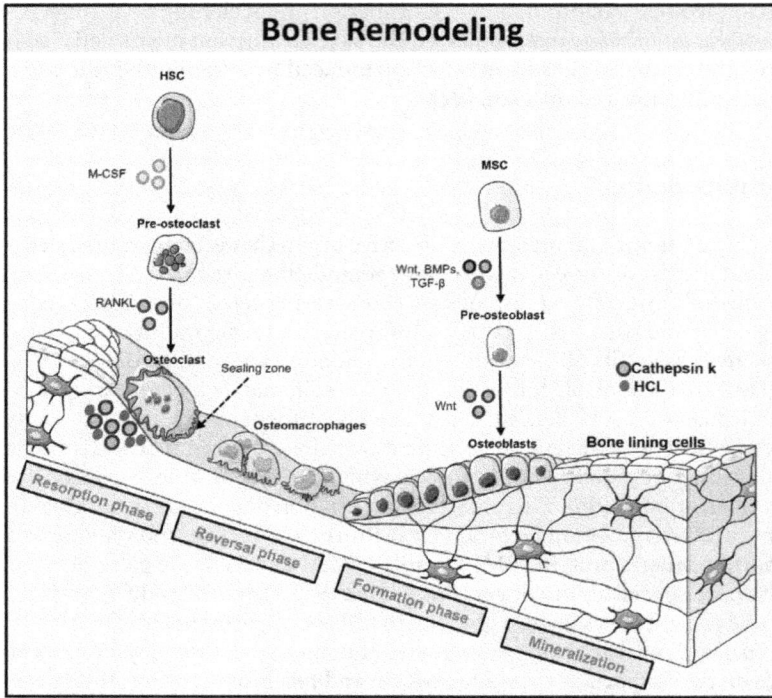

**Figure 2.**
*Schematic representation of bone remodeling. Bone remodeling occurs in four phases viz. 1) Activation phase: MCSF and RANKL induce the differentiation of osteoclast progenitors into osteoclasts. 2) Resorption phase: Mature osteoclast with unique ruffled border starts resorption of bone by secreting cathepsin K, and H⁺ in sealing zone. After resorption osteoclasts detach from the surface of bone and undergo apoptosis. 3) Reversal phase: During reversal phase osteoblasts precursor get differentiated into mature osteoblasts and are recruited to the resorption site. 4) Formation phase: Osteoblasts get occupied in the resorbed lacuna and start depositing the bone matrix. After formation phase osteoid gets mineralized and bone surface returns to resting phase with bone lining cells.*

the surface for osteoblasts where they can next start the process of bone formation. Mononuclear cell also provides various signals for the differentiation and migration of osteoblasts [93, 97]. During formation phase osteoblasts replace the resorbed bone with new bone [97] (**Figure 2**). Bone remodeling is regulated by various factors such as hormones like estrogen and parathyroid hormone and immune cells like T cells and B cells. Below we next discusse the role of immune system in regulation of bone health and the potential of LA in preventing bone resorption via immunomodulation.

## 4. Bone and immune system

Bone is an immunomodulatory organ and various immune cells affect the development of bone. Bone cells and immune cells interact with each other in the bone marrow which is the common niche for the development of both bone and immune system. In bone marrow, bone cells and immune cells interact with each other and affects each other development. The interaction between the bone and immune system is now studied under a new field of immunology termed as Osteoimmunology, a term coined by Choi et al. in 2006. Impact of various immune cells and cytokines secreted by immune cells on bone development is now known [97]. It is observed that cytokines such as IL-1, TNF-α, IL-6, IL-11, IL-15 and IL-17 induce bone resorption whereas cytokines such as IL-4, IL-10, IL13, IL-18, IFN-γ and granulocyte macrophage colony stimulating factor (GM-CSF) prevent bone loss. In various bone disorders the role of immune system has been discovered such as osteoporosis, RA and periodontitis. Osteoporosis is an inflammatory disease and several immune cells affect the development of osteoporosis. To study the immunology of osteoporosis we integrative biology started a novel field termed by us as "Immunoporosis" which deals specifically with the role of immune cells in osteoporosis [96]. Th17 cells and Tregs have most vital role in the development of bones and their balance is required for proper regulation of bone mass. CD4$^+$FOXP3$^+$ Tregs enhance bone mass by inhibiting osteoclastogenesis by directly suppressing the production of RANKL and MCSF [98]. Another mechanism by which CD4$^+$FOXP3$^+$ Tregs inhibit osteoclastogenesis or bone loss is by interacting with the CD80 and CD86 present on osteoclast precursors via CTLA-4, thereby inhibiting osteoclast differentiation [96]. Not only CD4$^+$FOXP3$^+$ Tregs, now the effect of CD8$^+$FOXP3$^+$ Tregs on bone is also discovered. It is observed that the CD8$^+$ Tregs prevent bone loss by inhibiting the formation of actin ring resulting in suppression of osteoclastogenesis [99]. Unlike Tregs, Th17 cells promote bone loss by inducing osteoclastogenesis via secretion of RANKL. Th17 also secrete IL-17 which induce bone loss by promoting RANKL expression on osteoclastogenesis supporting cells and by stimulating expression of inflammatory cytokines such as TNFα, IL-1 and IL-6 which further upregulate RANKL expression [9, 100]. Imbalance of Tregs and Th17 cells leads to bone loss which occurs during post-menopausal osteoporosis (PMO). Lack of estrogen promotes PMO. Estrogen prevents osteoporosis by inhibiting osteoclastogenesis but estrogen deficiency causes increased osteoclastogenesis by stimulating differentiation of Th17 cells. We also found in our studies that level of Th17 cells and inflammatory cytokines such as TNFα, IL-6, IL-17 and RANKL increased during post-menopausal osteoporosis [9, 101, 102]. Several studies have shown the role of Tregs and Th17 cells imbalanc in pathogenesis of RA [103]. The frequency of Th17 cells are enhanced in the joints and synovial fluid of RA patients [104] whereas the percentage of Tregs get significantly decreased in RA patients [105]. Similarly, the role of Tregs and Th17 cells imbalance is also found to be associated with periodontitis inflammation [106] and osteoarthritis [107]. Apart from these various other immune cells such as Th1, Th2, Th9 cells and γδ T cells are also involved in regulating bone health [97].

## 4.1 Role of LA in regulation of osteoimmune system

As immune system has such an important role in regulation of bone health, proper maintenance of immune homeostasis is very much required. Immune homeostasis for bone regulation is maintained by various factors such as estrogen hormone. Various strategies are used to prevent bone loss due to immune disruptions such as Denosumab, rituximab and TNF blockers [108, 109]. These strategies are proven effective but they also exert various adverse effects in the long run. Recently the use of probiotics is found to be effective in treatment of various inflammatory disorders such as IBD, obesity, diabetes, etc. [110–112]. Probiotics are also considered for the treatment of various bone disorders. LA is one of these probiotics. It is observed that LA has great potential of treating various bone pathologies. From comparison of different *Lactobacillus* species, it is observed that the effect of *Lactobacillus* on bone health is species dependent and LA has showed the most significant effect on bone parameters such as bone mineral density (BMD) and bone mineral content (BMC) among other *Lactobacillus* species [113]. In rat model of apical periodontitis, it is observed that level of alkaline phosphatase is significantly higher whereas the

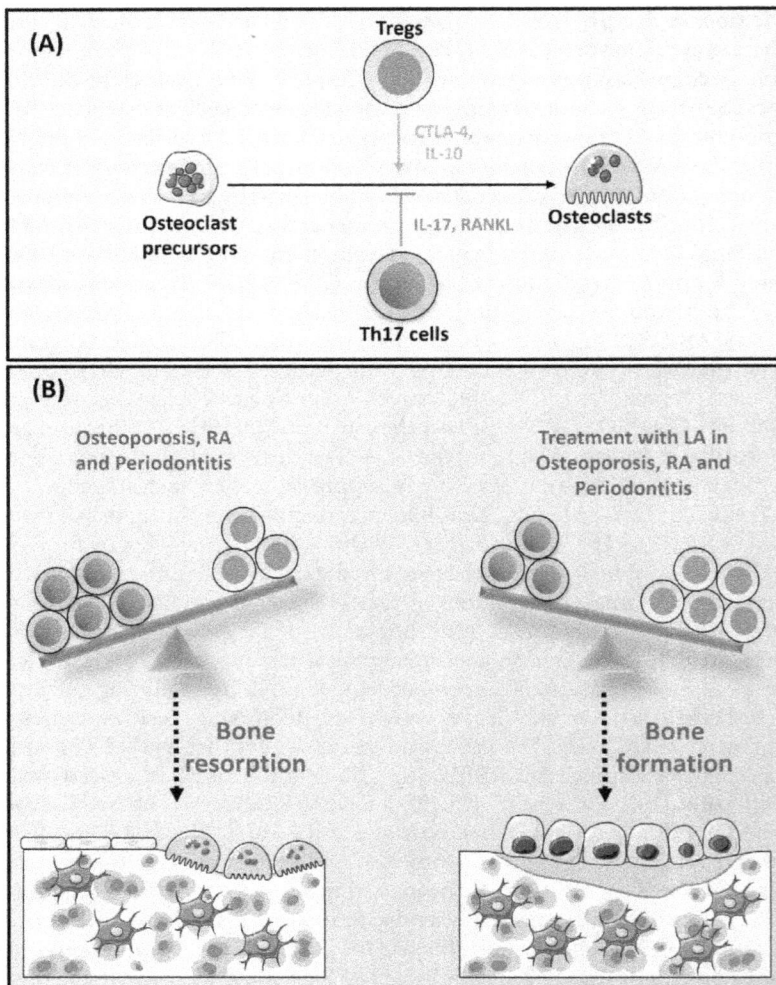

**Figure 3.**
*Role of Tregs/Th17 cells axis in regulation of bone health: (**A**) Tregs inhibit the differentiation of osteoclasts by secreting IL-10. Tregs also suppress osteoclastogenesis or bone loss by interacting with the CD80 and CD86 present on osteoclast precursors through cytotoxic T lymphocyte associated antigen 4 (CTLA-4). Th17 promote osteoclastogenesis via secretion of RANKL and IL-17. IL-17 induce expression of RANKL on osteoclastogenesis promoting cells. (**B**) In normal healthy conditions there is balance between Tregs and Th17 cells but during osteoporosis or other bone diseases like RA and periodontitis number of Th17 cells is increased which further leads to bone loss. LA treatment in these diseases restores the balance of Tregs and Th17 cells and thus prevent bone resorption.*

level of TRAP and RANKL is significantly lower in LA consumed groups [114]. It is reported that LA has antiarthritic properties and prevented Fruend's complete adjuvant mediated arthritis in female wistar rats [115]. It is observed that LA supernatant increased the proliferation of bone marrow stromal cells derived from rats [116]. It is observed in study from our group that LA can prevent bone loss by modulating the host immune system. We reported that LA improved both cortical and trabecular bone microarchitecture as well as enhanced the BMD and heterogeneity of bone in ovariectomized mice by skewing the Treg-Th17 cell balance (**Figure 3**). LA administration promoted the development of anti-osteoclastogenic Tregs and inhibited the osteoclastogenic Th17 cells in ovariectomized mice. LA supplementation also

| S.No. | Commercially available strains of LA | Source | Effect on bone | Reference |
|---|---|---|---|---|
| 1. | ATCC 4356 | ATCC | Modulated Treg-Th17 cell axis and inhibited the expression of inflammatory cytokines | [9] |
| 2. | ATCC 314 | ATCC | Prevented freund's complete adjuvant induced arthritis by decreasing the oxidative stress | [115, 118] |
| 2. | ATCC 11975 | ATCC | NR | — |
| 3. | ATCC 4375D-5 | ATCC | NR | — |
| 4. | ATCC 53671 | ATCC | NR | — |
| 5. | ATCC 4355 | ATCC | NR | — |
| 6. | ATCC 4357 | ATCC | NR | — |
| 7. | ATCC 9224 | ATCC | NR | — |
| 8. | ATCC BAA-2832 | ATCC | NR | — |
| 9. | ATCC 13651 | ATCC | NR | — |
| 10. | ATCC 11975 | ATCC | NR | — |
| 11. | ATCC 832 | ATCC | NR | — |
| 12. | ATCC 43121 | ATCC | NR | — |
| 13. | ATCC 53544 | ATCC | NR | — |
| 14. | ATCC 53545 | ATCC | NR | — |
| 15. | ATCC 53546 | ATCC | NR | — |
| 16. | ATCC 4796 | ATCC | NR | — |
| 18. | ATCC 53671 | ATCC | NR | — |
| 19. | ATCC 700396 | ATCC | NR | — |
| 20. | LA-1 | Chr. Hansen, Demark | Decreased the levels of inflammatory cytokines and enhanced the levels of anti-inflammatory cytokines in joints of osteoarthritic rats | [117] |
| 21. | LA-2 | Chr. Hansen, Demark | NR | — |
| 22 | LA-3 | Chr. Hansen, Demark | NR | — |
| 23 | LA-4 | Chr. Hansen, Demark | NR | — |
| 24 | LA-5 | Chr. Hansen, Demark | NR | — |
| 25 | LA-14 | Chr. Hansen, Demark | Decreased the inflammatory cytokines IL-1β and IL-6 in experimental apical periodontitis | [119] |
| 26 | DDS-1 | Nebraska cultures, Nebraska | NR | — |
| 27 | NCFM | Dansico, Madison | NR | — |
| 28 | SBT-2026 | Snow brand milk products, Japan | NR | — |

*ATCC: American Tissue Culture Collection.*
*NR: Not reported.*

**Table 1.**
*Different strains of LA and their effects on bone.*

attenuated the expression of osteoclastogenic cytokines such as IL-6, IL-17, RANKL, TNF-α and increased the expression of anti-osteoclastogenic cytokines like IL-10 and IFN-γ. Thus, LA has therapeutic effects and it can be used as an osteoprotective agent [9]. LA prevented monosodium iodoacetate induced osteoarthritis and reduced cartilage destruction via inhibition of proinflammatory cytokines production [117]. LA supplementation along with *L. rhamnosus* significantly decreased the inflammatory cytokines IL-1β and IL-6 and enhanced the expression of IL-10 as compared to the controls in experimental apical periodontitis [114]. LA supplementation upregulated anti-inflammatory cytokines and downregulated inflammatory cytokines in serum in experimental arthritis model [118]. Thus, LA has great ability of preventing inflammatory bone loss and of regulating osteoimmune system (**Table 1**).

## 5. Bone and dysbiosis

A number of microbes are localized in the gut. Some of them are beneficial for health whereas others are pathogenic and a balance of these microbes is required for normal physiological functioning of body. But due to several reasons like surgery, medications, irradiation and antibiotics this balance is dysregulated which leads to modifications in gut microbiota composition [3]. Dysbiosis is observed in various bone pathologies. Normally gut is dominated by four types of microbial phyla: *Firmicutes*, *Bacteroidetes*, *Proteobacteria* and *Actinobacteria*. *Firmicutes* and *Bacteroidetes* constitutes over 90% of the total gut microbiota and dysregulation of *Firmicutes/Bacteroidetes* ratio affect various biological processes like bone remodeling. During osteoporosis *Firmicutes* counts significantly increases whereas counts of *Bacteroidetes* significantly decreases [120–122]. In osteoporosis increase in the number of *Faecalibacterium* and *Dialister* genera is also reported [123]. Dysbiosis is also observed in RA and periodontitis [124, 125]. Although, LA mainly prevents bone loss by regulating immune homeostasis it also restores the gut microbiota composition in various diseases as discussed above. Thus, it can be possible that LA can also inhibit bone resorption by preventing dysbiosis.

## 6. Bone health and gut permeability

Gut permeability has very important role in regulation of bone health. Various studies have shown that increase in gut permeability is associated with bone loss. Collins et al. group measured the intestinal permeability after 1, 4 and 8 weeks of ovx surgery and they found increased intestinal permeability one week after ovariectomy along with the increase in inflammatory cytokines like IL-1β and TNFα which are responsible for bone loss [126]. Estrogen deficiency during postmenopausal osteoporosis is responsible for increase in gut permeability. Estrogen has very significant role in regulating the gut barrier. It maintains gut barrier through its receptors which are present on the intestinal epithelial cells. There are two types of estrogen receptors: ERα and ERβ. ERβ has very important role in the regulation of TJs as ERβ$^{-/-}$ mice has disrupted expression of tight junction proteins [39]. Various other studies proved the role of estrogen in regulation of gut barrier. Langen et al. reported decreased expression of ERβ and increased gut permeability in IBD patients [127]. Gut permeability decreases during oestrous phase whereas it is increased during dioestrus phase of rats and this increase in intestinal permeability during dioestrus phase can be prevented by treatment with oestradiol which upregulate the expression of occludin [128]. Estrogen and progesterone treatment

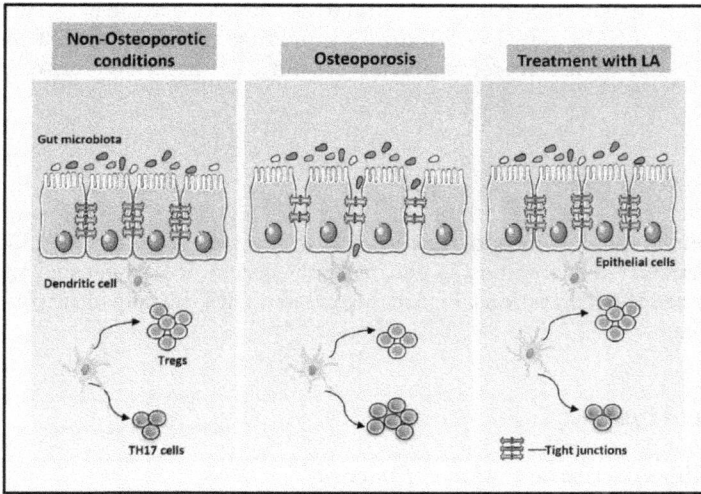

**Figure 4.**
*Proposed mechanism of LA in bone health. In normal healthy condition there is no dysbiosis and no alteration in gut permeability which prevents gut inflammation and thus bone loss. During osteoporosis gut permeability increases which leads to dysbiosis and bone loss. LA regulate Tregs/Th17 cell axis, prevents dysbiosis and maintains gut permeability. Altogether, LA treatment prevents leaky gut and dysbiosis thereby restoring gut immune homeostasis in osteoporosis which inhibit bone loss.*

decreases gut permeability and thus prevent secretion of inflammatory cytokines in IBD models [129]. LA prevents leaky gut in various diseases as discussed above and thus it can be possible that LA is effective in preventing leaky gut induced bone loss also. In summary we can say that by maintaining immune homeostasis and regulating both gut permeability and dysbiosis LA prevents bone resorption (**Figure 4**).

## 7. Conclusion

In the last few years, several studies have delineated the role of LA in preventing a number of inflammatory and metabolic disorders. LA prevents these disorders through various mechanism such as by modulating the host immune system, by maintaining the gut permeability along with preventing dysbiosis. The role of LA in suppressing bone loss also highlights the importance of LA in regulating bone health. LA enhances bone mass and prevents several bone diseases like osteoporosis, arthritis and periodontitis via regulating the immune homeostasis. Thus, immunomodulatory property of LA is of utmost importance in management of various bone pathologies. LA can also prevent bone resorption by regulating the leaky gut and dysbiosis. Thus, LA has immense potential as a probiotic and can be used as a medical therapy for treatment of bone loss in humans but before that a lot of research is further needed to be done on the efficacy along with the associated pros and cons of LA on human health.

## Acknowledgements

This work was financially supported by projects: DST-SERB (EMR/2016/007158), Govt. of India, Intramural project from All India Institute of Medical Sciences (A-798), New Delhi-India and AIIMS-IITD collaborative project

(AI-15) sanctioned to RKS. AB, LS, BV and RKS acknowledge the Department of Biotechnology, AIIMS, New Delhi-India for providing infrastructural facilities. AB thank DST SERB project for research fellowship, LS thank UGC for research fellowship.

## Author contributions

RKS contributed in conceptualization and writing of the manuscript. AB, LS and BV participated in writing and editing of the review. RKS suggested and AB and LS created the illustrations. Figures are created with the help of https://smart.servier.com.

## Conflicts of interest

The authors declare no conflicts of interest.

## Author details

Asha Bhardwaj, Leena Sapra, Bhupendra Verma and Rupesh K. Srivastava*
Department of Biotechnology, All India Institute of Medical Sciences (AIIMS), New Delhi, India

*Address all correspondence to: rupesh_srivastava13@yahoo.co.in; rupeshk@aiims.edu

IntechOpen

# References

[1] Ozen M, Dinleyici EC. The history of probiotics: the untold story. Benef Microbes. 2015 Jan 1;6(2):159-65.

[2] Gasbarrini G, Bonvicini F, Gramenzi A. Probiotics History. J Clin Gastroenterol. 2016 Nov;50(Supplement 2):S116-9.

[3] Williams NT. Probiotics. Am J Heal Pharm AJHP Off J Am Soc Heal Pharm. 2010 Mar;67(6):449-58.

[4] Gupta V, Garg R. Probiotics. Indian J Med Microbiol. 2009;27(3):202.

[5] Verna EC, Lucak S. Use of probiotics in gastrointestinal disorders: what to recommend? Therap Adv Gastroenterol. 2010 Sep 20;3(5):307-19.

[6] Plaza-Diaz J, Ruiz-Ojeda FJ, Gil-Campos M, Gil A. Mechanisms of Action of Probiotics. Adv Nutr. 2019 Jan 1;10(suppl_1):S49-66.

[7] Suez J, Zmora N, Segal E, Elinav E. The pros, cons, and many unknowns of probiotics. Nat Med. 2019 May 6;25(5):716-29.

[8] Sanders ME, Klaenhammer TR. Invited Review: The Scientific Basis of Lactobacillus acidophilus NCFM Functionality as a Probiotic. J Dairy Sci. 2001;84(2):319-31.

[9] Dar HY, Shukla P, Mishra PK, Anupam R, Mondal RK, Tomar GB, et al. Lactobacillus acidophilus inhibits bone loss and increases bone heterogeneity in osteoporotic mice via modulating Treg-Th17 cell balance. Bone reports. 2018 Jun;8:46-56.

[10] Bull M, Plummer S, Marchesi J, Mahenthiralingam E. The life history of *Lactobacillus acidophilus* as a probiotic: a tale of revisionary taxonomy, misidentification and commercial success. FEMS Microbiol Lett. 2013 Dec 1;349(2):77-87.

[11] Chen L, Zou Y, Peng J, Lu F, Yin Y, Li F, et al. Lactobacillus acidophilus suppresses colitis-associated activation of the IL-23/Th17 axis. J Immunol Res. 2015;2015:909514.

[12] Zhuo Q, Yu B, Zhou J, Zhang J, Zhang R, Xie J, et al. Lysates of Lactobacillus acidophilus combined with CTLA-4-blocking antibodies enhance antitumor immunity in a mouse colon cancer model. Sci Rep. 2019;9(1):20128.

[13] Nuraida L. A review: Health promoting lactic acid bacteria in traditional Indonesian fermented foods. Vol. 4, Food Science and Human Wellness. Elsevier B.V.; 2015. p. 47-55.

[14] Carr FJ, Chill D, Maida N. The lactic acid bacteria: A literature survey. Vol. 28, Critical Reviews in Microbiology. CRC Press LLC; 2002. p. 281-370.

[15] Arumugam M, Raes J, Pelletier E, Le Paslier D, Yamada T, Mende DR, et al. Enterotypes of the human gut microbiome. Nature. 2011 May 12;473(7346):174-80.

[16] Anjum N, Maqsood S, Masud T, Ahmad A, Sohail A, Momin A. Lactobacillus acidophilus: Characterization of the Species and Application in Food Production. Crit Rev Food Sci Nutr. 2014 Jan 5;54(9):1241-51.

[17] Claesson MJ, van Sinderen D, O'Toole PW. Lactobacillus phylogenomics - Towards a reclassification of the genus. Int J Syst Evol Microbiol. 2008;58(12):2945-54.

[18] Slover CM. Lactobacillus: a Review. Clin Microbiol Newsl. 2008 Feb 15;30(4):23-7.

[19] María Remes Troche J, Coss Adame E, Ángel Valdovinos Díaz M,

Gómez Escudero O, Eugenia Icaza Chávez M, Antonio Chávez-Barrera J, et al. Lactobacillus acidophilus LB: a useful pharmabiotic for the treatment of digestive disorders. Therap Adv Gastroenterol. 2020;13:1756284820971201.

[20] Wagner RD, Pierson C, Warner T, Dohnalek M, Hilty M, Balish E. Probiotic Effects of Feeding Heat-Killed Lactobacillus acidophilus and Lactobacillus casei to *Candida albicans*-Colonized Immunodeficient Mice. J Food Prot. 2000 May 1;63(5):638-44.

[21] Simakachorn N, Pichaipat V, Rithipornpaisarn P, Kongkaew C, Tongpradit P, Varavithya W. Clinical evaluation of the addition of lyophilized, heat-killed Lactobacillus acidophilus LB to oral rehydration therapy in the treatment of acute diarrhea in children. J Pediatr Gastroenterol Nutr. 2000 Jan;30(1):68-72.

[22] Foysal MJ, Fotedar R, Siddik MAB, Tay A. Lactobacillus acidophilus and *L. plantarum* improve health status, modulate gut microbiota and innate immune response of marron (Cherax cainii). Sci Rep. 2020;10(1):5916.

[23] Kaushal D, Kansal VK. Age-related decline in macrophage and lymphocyte functions in mice and its alleviation by treatment with probiotic Dahi containing Lactobacillus acidophilus and Bifidobacterium bifidum. J Dairy Res. 2011 Nov;78(4):404-11.

[24] Li S-C, Hsu W-F, Chang J-S, Shih C-K. Combination of Lactobacillus acidophilus and Bifidobacterium animalis subsp. lactis Shows a Stronger Anti-Inflammatory Effect than Individual Strains in HT-29 Cells. Nutrients. 2019 Apr 27;11(5).

[25] Perdigón G, de Macias ME, Alvarez S, Oliver G, de Ruiz Holgado AP.

Systemic augmentation of the immune response in mice by feeding fermented milks with Lactobacillus casei and Lactobacillus acidophilus. Immunology. 1988 Jan;63(1):17-23.

[26] Kim Y, Mylonakis E. Caenorhabditis elegans immune conditioning with the probiotic bacterium Lactobacillus acidophilus strain NCFM enhances gram-positive immune responses. Infect Immun. 2012 Jul;80(7):2500-8.

[27] Na HG, Park Y, Kim M-A, Lee JW, So G, Kim SH, et al. Secondary Fermented Extract of Chaga-Cheonggukjang Attenuates the Effects of Obesity and Suppresses Inflammatory Response in the Liver and Spleen of High-Fat Diet-Induced Obese Mice. J Microbiol Biotechnol. 2019 May 28;29(5):739-48.

[28] Neumann E, Oliveira MA, Cabral CM, Moura LN, Nicoli JR, Vieira EC, et al. Monoassociation with Lactobacillus acidophilus UFV-H2b20 stimulates the immune defense mechanisms of germfree mice. Brazilian J Med Biol Res = Rev Bras Pesqui medicas e Biol. 1998 Dec;31(12):1565-73.

[29] Azevedo MSP, Zhang W, Wen K, Gonzalez AM, Saif LJ, Yousef AE, et al. Lactobacillus acidophilus and *Lactobacillus reuteri* modulate cytokine responses in gnotobiotic pigs infected with human rotavirus. Benef Microbes. 2012 Mar 1;3(1):33-42.

[30] Vissers YM, Snel J, Zuurendonk PF, Smit BA, Wichers HJ, Savelkoul HFJ. Differential effects of Lactobacillus acidophilus and *Lactobacillus plantarum* strains on cytokine induction in human peripheral blood mononuclear cells. FEMS Immunol Med Microbiol. 2010 Jun 1;59(1):60-70.

[31] Magistrelli L, Amoruso A, Mogna L, Graziano T, Cantello R, Pane M, et al. Probiotics May Have Beneficial Effects in Parkinson's Disease: In vitro Evidence. Front Immunol. 2019;10:969.

[32] Fukuyama K, Islam MA, Takagi M, Ikeda-Ohtsubo W, Kurata S, Aso H, et al. Evaluation of the Immunomodulatory Ability of Lactic Acid Bacteria Isolated from Feedlot Cattle Against Mastitis Using a Bovine Mammary Epithelial Cells In Vitro Assay. Pathogens. 2020 May 25;9(5):410.

[33] Shojadoost B, Kulkarni RR, Brisbin JT, Quinteiro-Filho W, Alkie TN, Sharif S. Interactions between lactobacilli and chicken macrophages induce antiviral responses against avian influenza virus. Res Vet Sci. 2019 Aug;125:441-50.

[34] Kim JY. Probiotic modulation of dendritic cells co-cultured with intestinal epithelial cells. World J Gastroenterol. 2012;18(12):1308.

[35] Steinberg RS, Lima M, Gomes de Oliveira NL, Miyoshi A, Nicoli JR, Neumann E, et al. Effect of intestinal colonisation by two Lactobacillus strains on the immune response of gnotobiotic mice. Benef Microbes. 2014 Dec;5(4):409-19.

[36] Park J-S, Choi JW, Jhun J, Kwon JY, Lee B-I, Yang CW, et al. Lactobacillus acidophilus Improves Intestinal Inflammation in an Acute Colitis Mouse Model by Regulation of Th17 and Treg Cell Balance and Fibrosis Development. J Med Food. 2018 Mar;21(3):215-24.

[37] Kim W-K, Han DH, Jang YJ, Park S, Jang SJ, Lee G, et al. Alleviation of DSS-induced colitis via Lactobacillus acidophilus treatment in mice. Food Funct. 2021 Jan 7;12(1):340-50.

[38] Kim DH, Kim S, Lee JH, Kim JH, Che X, Ma HW, et al. Lactobacillus acidophilus suppresses intestinal inflammation by inhibiting endoplasmic reticulum stress. J Gastroenterol Hepatol. 2019 Jan;34(1):178-85.

[39] Goto H, Sagitani A, Ashida N, Kato S, Hirota T, Shinoda T,

et al. Anti-influenza virus effects of both live and non-live Lactobacillus acidophilus L-92 accompanied by the activation of innate immunity. Br J Nutr. 2013 Nov;110(10):1810-8.

[40] Wang Y, Gu Y, Fang K, Mao K, Dou J, Fan H, et al. Lactobacillus acidophilus and Clostridium butyricum ameliorate colitis in murine by strengthening the gut barrier function and decreasing inflammatory factors. Benef Microbes. 2018 Sep 18;9(5):775-87.

[41] Wang J-J, Zhang Q-M, Ni W-W, Zhang X, Li Y, Li A-L, et al. Modulatory effect of Lactobacillus acidophilus KLDS 1.0738 on intestinal short-chain fatty acids metabolism and GPR41/43 expression in β-lactoglobulin-sensitized mice. Microbiol Immunol. 2019 Aug;63(8):303-15.

[42] Wen K, Li G, Bui T, Liu F, Li Y, Kocher J, et al. High dose and low dose Lactobacillus acidophilus exerted differential immune modulating effects on T cell immune responses induced by an oral human rotavirus vaccine in gnotobiotic pigs. Vaccine. 2012 Feb 1;30(6):1198-207.

[43] Shah MM, Saio M, Yamashita H, Tanaka H, Takami T, Ezaki T, et al. Lactobacillus acidophilus strain L-92 induces CD4(+)CD25(+)Foxp3(+) regulatory T cells and suppresses allergic contact dermatitis. Biol Pharm Bull. 2012;35(4):612-6.

[44] Roselli M, Finamore A, Nuccitelli S, Carnevali P, Brigidi P, Vitali B, et al. Prevention of TNBS-induced colitis by different Lactobacillus and Bifidobacterium strains is associated with an expansion of gammadeltaT and regulatory T cells of intestinal intraepithelial lymphocytes. Inflamm Bowel Dis. 2009 Oct;15(10):1526-36.

[45] Li A-L, Sun Y-Q, Du P, Meng X-C, Guo L, Li S, et al. The Effect of

Lactobacillus actobacillus Peptidoglycan on Bovine β-Lactoglobulin-Sensitized Mice via TLR2/NF-κB Pathway. Iran J Allergy Asthma Immunol. 2017 Apr;16(2):147-58.

[46] Paturi G, Phillips M, Kailasapathy K. Effect of probiotic strains Lactobacillus acidophilus LAFTI L10 and *Lactobacillus paracasei* LAFTI L26 on systemic immune functions and bacterial translocation in mice. J Food Prot. 2008 Apr;71(4):796-801.

[47] Zhang D, Wu M, Guo Y, Xun M, Wang W, Wu Z, et al. Purification of Lactobacillus acidophilus surface-layer protein and its immunomodulatory effects on RAW264.7 cells. J Sci Food Agric. 2017 Sep 1;97(12):4204-9.

[48] Gill HS, Rutherfurd KJ, Prasad J, Gopal PK. Enhancement of natural and acquired immunity by *Lactobacillus rhamnosus* (HN001), Lactobacillus acidophilus (HN017) and Bifidobacterium lactis (HN019). Br J Nutr. 2000 Feb;83(2):167-76.

[49] Rangavajhyala N, Shahani KM, Sridevi G, Srikumaran S. Nonlipopolysaccharide components) of Lactobacillus addophilus stimulate(s) the production of interleukin-1α and tumor necrosis factor-α by murine macrophages. Nutr Cancer. 1997 Jan;28(2):130-4.

[50] Elawadli I, Brisbin JT, Mallard BA, Griffiths MW, Corredig M, Sharif S. Differential effects of lactobacilli on activation and maturation of mouse dendritic cells. Benef Microbes. 2014 Sep;5(3):323-34.

[51] Weiss G, Maaetoft-Udsen K, Stifter SA, Hertzog P, Goriely S, Thomsen AR, et al. MyD88 drives the IFN-β response to Lactobacillus acidophilus in dendritic cells through a mechanism involving IRF1, IRF3, and IRF7. J Immunol. 2012 Sep 15;189(6):2860-8.

[52] Konstantinov SR, Smidt H, de Vos WM, Bruijns SCM, Singh SK, Valence F, et al. S layer protein A of Lactobacillus acidophilus NCFM regulates immature dendritic cell and T cell functions. Proc Natl Acad Sci U S A. 2008 Dec 9;105(49):19474-9.

[53] Kim JY, Choi YO, Ji GE. Effect of oral probiotics (Bifidobacterium lactis AD011 and Lactobacillus acidophilus AD031) administration on ovalbumin-induced food allergy mouse model. J Microbiol Biotechnol. 2008 Aug;18(8):1393-400.

[54] Cheon S, Lee KW, Kim KE, Park JK, Park S, Kim C, et al. Heat-killed Lactobacillus acidophilus La205 enhances NK cell cytotoxicity through increased granule exocytosis. Immunol Lett. 2011 May;136(2):171-6.

[55] De Simone C, Ciardi A, Grassi A, Lambert Gardini S, Tzantzoglou S, Trinchieri V, et al. Effect of Bifidobacterium bifidum and Lactobacillus acidophilus on gut mucosa and peripheral blood B lymphocytes. Immunopharmacol Immunotoxicol. 1992;14(1-2):331-40.

[56] Li A-L, Meng X-C, Duan C-C, Huo G-C, Zheng Q-L, Li D. Suppressive effects of oral administration of heat-killed Lactobacillus acidophilus on T helper-17 immune responses in a bovine β-lactoglobulin-sensitized mice model. Biol Pharm Bull. 2013;36(2):202-7.

[57] Liu F, Wen K, Li G, Yang X, Kocher J, Bui T, et al. Dual functions of Lactobacillus acidophilus NCFM as protection against rotavirus diarrhea. J Pediatr Gastroenterol Nutr. 2014 Feb;58(2):169-76.

[58] Paturi G, Phillips M, Jones M, Kailasapathy K. Immune enhancing effects of Lactobacillus acidophilus LAFTI L10 and *Lactobacillus paracasei* LAFTI L26 in mice. Int J Food Microbiol. 2007 Apr 1;115(1):115-8.

[59] Nouri Gharajalar S, Mirzai P, Nofouzi K, Madadi MS. Immune enhancing effects of Lactobacillus acidophilus on Newcastle disease vaccination in chickens. Comp Immunol Microbiol Infect Dis. 2020 Oct;72:101520.

[60] Jain S, Yadav H, Sinha PR, Naito Y, Marotta F. Dahi containing probiotic Lactobacillus acidophilus and Lactobacillus casei has a protective effect against Salmonella enteritidis infection in mice. Int J Immunopathol Pharmacol. 21(4):1021-9.

[61] Su J, Li J, Zheng H, You Y, Luo X, Li Y, et al. Adjuvant effects of L. acidophilus LW1 on immune responses to the foot-and-mouth disease virus DNA vaccine in mice. PLoS One. 2014;9(8):e104446.

[62] Li C-Y, Lin H-C, Lai C-H, Lu JJ-Y, Wu S-F, Fang S-H. Immunomodulatory effects of lactobacillus and Bifidobacterium on both murine and human mitogen-activated T cells. Int Arch Allergy Immunol. 2011;156(2):128-36.

[63] Carding S, Verbeke K, Vipond DT, Corfe BM, Owen LJ. Dysbiosis of the gut microbiota in disease. Microb Ecol Heal Dis. 2015 Feb 2;26.

[64] Chong PP, Chin VK, Looi CY, Wong WF, Madhavan P, Yong VC. The Microbiome and Irritable Bowel Syndrome – A Review on the Pathophysiology, Current Research and Future Therapy. Front Microbiol. 2019 Jun 10;10.

[65] Tamboli CP. Dysbiosis in inflammatory bowel disease. Gut. 2004 Jan 1;53(1):1-4.

[66] Kesh K, Mendez R, Abdelrahman L, Banerjee S, Banerjee S. Type 2 diabetes induced microbiome dysbiosis is associated with therapy resistance in pancreatic adenocarcinoma. Microb Cell Fact. 2020 Dec 24;19(1):75.

[67] Nagpal R, Newman TM, Wang S, Jain S, Lovato JF, Yadav H. Obesity-Linked Gut Microbiome Dysbiosis Associated with Derangements in Gut Permeability and Intestinal Cellular Homeostasis Independent of Diet. J Diabetes Res. 2018 Sep 3;2018:1-9.

[68] Bindels LB, Beck R, Schakman O, Martin JC, De Backer F, Sohet FM, et al. Restoring specific lactobacilli levels decreases inflammation and muscle atrophy markers in an acute leukemia mouse model. PLoS One. 2012;7(6):e37971.

[69] Russo R, Edu A, De Seta F. Study on the effects of an oral lactobacilli and lactoferrin complex in women with intermediate vaginal microbiota. Arch Gynecol Obstet. 2018;298(1):139-45.

[70] Salinas AM, Osorio VG, Endara PF, Salazar ER, Vasco GP, Vivero SG, et al. Bacterial identification of the vaginal microbiota in Ecuadorian pregnant teenagers: an exploratory analysis. PeerJ. 2018;6:e4317.

[71] Vemuri R, Gundamaraju R, Shinde T, Perera AP, Basheer W, Southam B, et al. Lactobacillus acidophilus DDS-1 Modulates Intestinal-Specific Microbiota, Short-Chain Fatty Acid and Immunological Profiles in Aging Mice. Nutrients. 2019 Jun 7;11(6).

[72] Palumbo VD, Romeo M, Marino Gammazza A, Carini F, Damiani P, Damiano G, et al. The long-term effects of probiotics in the therapy of ulcerative colitis: A clinical study. Biomed Pap Med Fac Univ Palacky Olomouc Czech Repub. 2016 Sep;160(3):372-7.

[73] Pradhan B, Guha D, Naik AK, Banerjee A, Tambat S, Chawla S, et al. Probiotics L. acidophilus and B. clausii Modulate Gut Microbiota in Th1- and Th2-Biased Mice to Ameliorate Salmonella Typhimurium-Induced Diarrhea. Probiotics Antimicrob Proteins. 2019;11(3):887-904.

[74] Bagarolli RA, Tobar N, Oliveira AG, Araújo TG, Carvalho BM, Rocha GZ, et al. Probiotics modulate gut microbiota and improve insulin sensitivity in DIO mice. J Nutr Biochem. 2017;50:16-25.

[75] Deol PK, Khare P, Singh DP, Soman G, Bishnoi M, Kondepudi KK, et al. Managing colonic inflammation associated gut derangements by systematically optimised and targeted ginger extract-Lactobacillus acidophilus loaded pharmacobiotic alginate beads. Int J Biol Macromol. 2017 Dec;105(Pt 1):81-91.

[76] Chen T, Xia C, Hu H, Wang H, Tan B, Tian P, et al. Dysbiosis of the rat vagina is efficiently rescued by vaginal microbiota transplantation or probiotic combination. Int J Antimicrob Agents. 2021 Jan 9;106277.

[77] Shoaib A, Dachang W, Xin Y. Determining the role of a probiotic in the restoration of intestinal microbial balance by molecular and cultural techniques. Genet Mol Res. 2015 Feb 20;14(1):1526-37.

[78] Konturek PC, Konturek K, Brzozowski T, Wojcik D, Magierowski M, Targosz A, et al. Participation of the intestinal microbiota in the mechanism of beneficial effect of treatment with synbiotic Syngut on experimental colitis under stress conditions. J Physiol Pharmacol. 2020 Jun;71(3).

[79] Kocián J. [Lactobacilli in the treatment of dyspepsia due to dysmicrobia of various causes]. Vnitr Lek. 1994 Feb;40(2):79-83.

[80] Li H, Shi J, Zhao L, Guan J, Liu F, Huo G, et al. *Lactobacillus plantarum* KLDS1.0344 and Lactobacillus acidophilus KLDS1.0901 Mixture Prevents Chronic Alcoholic Liver Injury in Mice by Protecting the Intestinal Barrier and Regulating Gut Microbiota and Liver-Related Pathways. J Agric Food Chem. 2021 Jan 13;69(1):183-97.

[81] Lee SH. Intestinal Permeability Regulation by Tight Junction: Implication on Inflammatory Bowel Diseases. Intest Res. 2015;13(1):11.

[82] Takeuchi K, Maiden L, Bjarnason I. Genetic aspects of intestinal permeability in inflammatory bowel disease. Novartis Found Symp. 2004;263:151-8; discussion 159-63, 211-8.

[83] Vu TH, Shipley JM, Bergers G, Berger JE, Helms JA, Hanahan D, et al. MMP-9/Gelatinase B Is a Key Regulator of Growth Plate Angiogenesis and Apoptosis of Hypertrophic Chondrocytes. Cell. 1998 May;93(3):411-22.

[84] Zeissig S, Burgel N, Gunzel D, Richter J, Mankertz J, Wahnschaffe U, et al. Changes in expression and distribution of claudin 2, 5 and 8 lead to discontinuous tight junctions and barrier dysfunction in active Crohn's disease. Gut. 2007 Jan 1;56(1):61-72.

[85] Edelblum KL, Turner JR. The tight junction in inflammatory disease: communication breakdown. Curr Opin Pharmacol. 2009 Dec;9(6):715-20.

[86] Vetrano S, Rescigno M, Rosaria Cera M, Correale C, Rumio C, Doni A, et al. Unique Role of Junctional Adhesion Molecule-A in Maintaining Mucosal Homeostasis in Inflammatory Bowel Disease. Gastroenterology. 2008 Jul;135(1):173-84.

[87] Soler AP. Increased tight junctional permeability is associated with the development of colon cancer. Carcinogenesis. 1999 Aug 1;20(8):1425-32.

[88] Bjarnason I, So A, Levi AJ, Peters T, Williams P, Zanelli G, et al. Intestinal permeability and inflammation in rheumatoid arthritis: Effects of non-steroidal anti-inflammatory drugs. Lancet. 1984 Nov;324(8413):1171-4.

[89] Guo S, Gillingham T, Guo Y, Meng D, Zhu W, Walker WA, et al. Secretions of Bifidobacterium infantis and Lactobacillus acidophilus Protect Intestinal Epithelial Barrier Function. J Pediatr Gastroenterol Nutr. 2017;64(3):404-12.

[90] Kim TK, Lee J-C, Im S-H, Lee M-S. Amelioration of Autoimmune Diabetes of NOD Mice by Immunomodulating Probiotics. Front Immunol. 2020;11:1832.

[91] Yu Y, Lu J, Oliphant K, Gupta N, Claud K, Lu L. Maternal administration of probiotics promotes gut development in mouse offsprings. Aguila MB, editor. PLoS One. 2020 Aug 7;15(8):e0237182.

[92] Resta-Lenert S, Barrett KE. Live probiotics protect intestinal epithelial cells from the effects of infection with enteroinvasive *Escherichia coli* (EIEC). Gut. 2003 Jul;52(7):988-97.

[93] Hadjidakis DJ, Androulakis II. Bone remodeling. In: Annals of the New York Academy of Sciences. Blackwell Publishing Inc.; 2006. p. 385-96.

[94] Walsh MC, Kim N, Kadono Y, Rho J, Lee SY, Lorenzo J, et al. Osteoimmunology: interplay between the immune system and bone metabolism. Annu Rev Immunol. 2006;24:33-63.

[95] Schett G, David J-P. The multiple faces of autoimmune-mediated bone loss. Nat Rev Endocrinol. 2010 Dec;6(12):698-706.

[96] Srivastava RK, Dar HY, Mishra PK. Immunoporosis: Immunology of Osteoporosis—Role of T Cells. Front Immunol. 2018;9:657.

[97] Dar HY, Azam Z, Anupam R, Mondal RK, Srivastava RK. Osteoimmunology: The Nexus between bone and immune system. Front Biosci (Landmark Ed. 2018 Jan 1;23:464-92.

[98] Zaiss MM, Axmann R, Zwerina J, Polzer K, Gückel E, Skapenko A, et al. Treg cells suppress osteoclast formation: A new link between the immune system and bone. Arthritis Rheum. 2007 Dec;56(12):4104-12.

[99] Shashkova E V., Trivedi J, Cline-Smith AB, Ferris C, Buchwald ZS, Gibbs J, et al. Osteoclast-Primed Foxp3 + CD8 T Cells Induce T-bet, Eomesodermin, and IFN-γ To Regulate Bone Resorption. J Immunol. 2016 Aug 1;197(3):726-35.

[100] Adamopoulos IE, Bowman EP. Immune regulation of bone loss by Th17 cells. Arthritis Res Ther. 2008;10(5):225.

[101] Dar HY, Pal S, Shukla P, Mishra PK, Tomar GB, Chattopadhyay N, et al. *Bacillus clausii* inhibits bone loss by skewing Treg-Th17 cell equilibrium in postmenopausal osteoporotic mice model. Nutrition. 2018;54:118-28.

[102] Sapra L, Dar HY, Bhardwaj A, Pandey A, Kumari S, Azam Z, et al. *Lactobacillus rhamnosus* attenuates bone loss and maintains bone health by skewing Treg-Th17 cell balance in Ovx mice. Sci Rep. 2021;11(1):1807.

[103] Kikodze N, Pantsulaia I, Chikovani T. The role of T regulatory and Th17 cells in the pathogenesis of rheumatoid arthritis (Review). Georgian Med News. 2016 Dec;(261):62-8.

[104] Leipe J, Grunke M, Dechant C, Reindl C, Kerzendorf U, Schulze-Koops H, et al. Role of Th17 cells in human autoimmune arthritis. Arthritis Rheum. 2010 Oct;62(10):2876-85.

[105] Zhang X, Zhang X, Zhuang L, Xu C, Li T, Zhang G, et al. Decreased regulatory T-cell frequency and interleukin-35 levels in patients with rheumatoid arthritis. Exp Ther Med. 2018 Oct 19;

[106] Gao L, Zhao Y, Wang P, Zhang L, Zhang C, Chen Q, et al. Detection of Th17/Treg cells and related factors in gingival tissues and peripheral blood of rats with experimental periodontitis. Iran J Basic Med Sci. 2017 Mar;20(3):294-300.

[107] Li Y, Luo W, Zhu S, Lei G. T Cells in Osteoarthritis: Alterations and Beyond. Front Immunol. 2017 Mar 30;8.

[108] Raterman HG, Lems WF. Pharmacological Management of Osteoporosis in Rheumatoid Arthritis Patients: A Review of the Literature and Practical Guide. Drugs Aging. 2019 Dec 21;36(12):1061-72.

[109] Elshahaly M, Wheater G, Naraghi K, Tuck SP, Datta HK, Ng W-F, et al. Changes in bone density and bone turnover in patients with rheumatoid arthritis treated with rituximab, a B cell depleting monoclonal antibody (HORUS TRIAL). BMC Musculoskelet Disord. 2013 Feb 14;14(S1):A10.

[110] Guandalini S, Sansotta N. Probiotics in the Treatment of Inflammatory Bowel Disease. Adv Exp Med Biol. 2019;1125:101-7.

[111] Wang Z-B, Xin S-S, Ding L-N, Ding W-Y, Hou Y-L, Liu C-Q, et al. The Potential Role of Probiotics in Controlling Overweight/Obesity and Associated Metabolic Parameters in Adults: A Systematic Review and Meta-Analysis. Evidence-Based Complement Altern Med. 2019 Apr 15;2019:1-14.

[112] Kocsis T, Molnár B, Németh D, Hegyi P, Szakács Z, Bálint A, et al. Probiotics have beneficial metabolic effects in patients with type 2 diabetes mellitus: a meta-analysis of randomized clinical trials. Sci Rep. 2020;10(1):11787.

[113] Montazeri-Najafabady N, Ghasemi Y, Dabbaghmanesh MH, Talezadeh P, Koohpeyma F, Gholami A. Supportive Role of Probiotic Strains in Protecting Rats from Ovariectomy-Induced Cortical Bone Loss. Probiotics Antimicrob Proteins. 2019;11(4):1145-54.

[114] Cosme-Silva L, Dal-Fabbro R, Cintra LTA, Dos Santos VR, Duque C, Ervolino E, et al. Systemic administration of probiotics reduces the severity of apical periodontitis. Int Endod J. 2019 Dec;52(12):1738-49.

[115] Amdekar S, Roy P, Singh V, Kumar A, Singh R, Sharma P. Anti-Inflammatory Activity of *Lactobacillus* on Carrageenan-Induced Paw Edema in Male Wistar Rats. Kanai T, editor. Int J Inflam. 2012;2012:752015.

[116] Samadikuchaksaraei A, Gholipourmalekabadi M, Saberian M, Abdollahpour Alitappeh M, Shahidi Delshad E. How does the supernatant of Lactobacillus acidophilus affect the proliferation and differentiation activities of rat bone marrow-derived stromal cells? Cell Mol Biol (Noisy-le-grand). 2016 Aug 31;62(10):1-6.

[117] Lee SH, Kwon JY, Jhun JY, Jung KA, Park SH, Yang CW, et al. Lactobacillus acidophilus ameliorates pain and cartilage degradation in experimental osteoarthritis. Immunol Lett. 2018 Nov 1;203:6-14.

[118] Amdekar S, Singh V, Kumar A, Sharma P, Singh R. Lactobacillus acidophilus Protected Organs in Experimental Arthritis by Regulating the Pro-inflammatory Cytokines. Indian J Clin Biochem. 2014 Oct;29(4):471-8.

[119] Cosme-Silva L, Dal-Fabbro R, Cintra LTA, Ervolino E, Plazza F, Mogami Bomfim S, et al. Reduced bone resorption and inflammation in apical periodontitis evoked by dietary supplementation with probiotics in rats. Int Endod J. 2020 Aug;53(8):1084-92.

[120] Wang J, Wang Y, Gao W, Wang B, Zhao H, Zeng Y, et al. Diversity analysis of gut microbiota in osteoporosis and osteopenia patients. PeerJ. 2017;5:e3450.

[121] Qin J, Li R, Raes J, Arumugam M, Burgdorf KS, Manichanh C, et al. A human gut microbial gene catalogue established by metagenomic sequencing. Nature. 2010 Mar;464(7285):59-65.

[122] Yatsonsky Ii D, Pan K, Shendge VB, Liu J, Ebraheim NA. Linkage of microbiota and osteoporosis: A mini literature review. World J Orthop. 2019 Mar 18;10(3):123-7.

[123] Xu Z, Xie Z, Sun J, Huang S, Chen Y, Li C, et al. Gut Microbiome Reveals Specific Dysbiosis in Primary Osteoporosis. Front Cell Infect Microbiol. 2020;10:160.

[124] Picchianti-Diamanti A, Panebianco C, Salemi S, Sorgi M, Di Rosa R, Tropea A, et al. Analysis of Gut Microbiota in Rheumatoid Arthritis Patients: Disease-Related Dysbiosis and Modifications Induced by Etanercept. Int J Mol Sci. 2018 Sep 27;19(10):2938.

[125] Nath S, Raveendran R. Microbial dysbiosis in periodontitis. J Indian Soc Periodontol. 2013;17(4):543.

[126] Collins FL, Rios-Arce ND, Atkinson S, Bierhalter H, Schoenherr D, Bazil JN, et al. Temporal and regional intestinal changes in permeability, tight junction, and cytokine gene expression following ovariectomy-induced estrogen deficiency. Physiol Rep. 2017 May;5(9):e13263.

[127] Looijer-van Langen M, Hotte N, Dieleman LA, Albert E, Mulder C, Madsen KL. Estrogen receptor-β signaling modulates epithelial barrier function. Am J Physiol Liver Physiol. 2011 Apr;300(4):G621-6.

[128] Braniste V, Leveque M, Buisson-Brenac C, Bueno L, Fioramonti J, Houdeau E. Oestradiol decreases colonic permeability through oestrogen receptor β-mediated up-regulation of occludin and junctional adhesion molecule-A in epithelial cells. J Physiol. 2009 Jul 1;587(13):3317-28.

[129] van der Giessen J, van der Woude CJ, Peppelenbosch MP, Fuhler GM. A Direct Effect of Sex Hormones on Epithelial Barrier Function in Inflammatory Bowel Disease Models. Cells. 2019;8(3).